IT'S A WORLD THING

Geography for Edexcel GCSE Specification B

Bob Digby Dan Cowling Jane Entwistle
Carole Goddard Peter Goddard Sue Warn

OXFORD UNIVERSITY PRESS

OXFORD
UNIVERSITY PRESS

Great Clarendon Street, Oxford OX2 6DP

Oxford University Press is a department of the University of Oxford. It furthers the University's objective of excellence in research, scholarship, and education by publishing worldwide in

Oxford New York

Athens Auckland Bangkok Bogotá Buenos Aires Cape Town
Chennai Dar es Salaam Delhi Florence Hong Kong Istanbul Karachi
Kolkata Kuala Lumpur Madrid Melbourne Mexico City Mumbai Nairobi
Paris São Paulo Shanghai Singapore Taipei Tokyo Toronto Warsaw

with associated companies in Berlin Ibadan

Oxford is a registered trade mark of Oxford University Press in the UK and in certain other countries

© Bob Digby, Dan Cowling, Jane Entwistle, Carole Goddard, Peter Goddard, Sue Warn 2001

The moral rights of the author have been asserted

Database right Oxford University Press (maker)

First published 2001

All rights reserved. No part of this publication may be reproduced, stored in a retrieval system, or transmitted, in any form or by any means, without the prior permission in writing of Oxford University Press, or as expressly permitted by law, or under terms agreed with the appropriate reprographics rights organization. Enquiries concerning reproduction outside the scope of the above should be sent to the Rights Department, Oxford University Press, at the address above.

You must not circulate this book in any other binding or cover and you must impose this same condition on any acquirer.

Third party website addresses referred to in this publication are provided by Oxford University Press in good faith and for information only and Oxford University Press disclaims any responsibility for the material contained therein.

British Library Cataloguing in Publication Data

Data available

ISBN 0 19 913428 6

Editorial, design, and production by Hart McLeod, Cambridge

Printed by Butler & Tanner Ltd, Frome, UK

Acknowledgements

The publisher and authors would like to thank the following for permission to use photographs and other copyright material (the first number gives the page, the second number gives the figure):

8.2 Countrywide Photographic, Courtesy of Ashford Borough Council; 10.1, 10.2, 11.3, 11.4, 12.2, 12.2 (bottom) Carole Goddard; 13.3 McArthur Glen; 14.1A Jean-Léo Dugast, Panos Pictures; 14.1B Dermot Tatlow, Panos Pictures; 15.3 Liba Taylor, Panos Pictures; 15.4 Chris Stowers, Panos Pictures; 17.3 Yann Arthus-Bertranc, Corbis; 18.2, 18.3, 19.4 Mark Edwards, Still Pictures; 19.5 Giacomo Pirozzi, Panos Pictures; 20.1 Ron Giling, Still Pictures; 20.2 Sean Sprague, Panos Pictures; 21.3 Paul Smith, Panos Pictures; 21.4 Maria Luiza M. Carvalho, Panos Pictures; 22.1 Alain le Garsmeur, Panos Pictures; 22.2 Still Pictures; 22.3 Gerard & Margi Moss, Still Pictures; 23.4 Mark Edwards, Still Pictures; 12.5 Mattews Foure, UNEP, Still Pictures; 24.1 Mark Edwards, Still Pictures; 24.3 Peter Currell, Ecoscene; 26.1 Associated Press AP; 27.3 Associated Press AP; 28 (bottom left) Mark Edwards, Still Pictures; 28 (bottom right) David Hoffman, Still Pictures; 29.4 Martin Jones, Ecoscene; 31.3 Action Aid, Agriculture and Development in Equatorial Uganda, 1997; 32.1 Adam Woolfitt, Corbis; 32.3 Associated Press AP; 33.5 Liba Taylor, Corbis; 34.1 (top right) Courtesy of Ford Motors, England; 34.1 (top left) Courtesy of Land Rover; 34.1 (bottom right) Courtesy of Jaguar; 34.1 (bottom left) Courtesy of Volvo; 36.1 Courtesy of Ford, India; 37.3 Courtesy of Ford, India; 39.3 Source: Washington Post, 1995; 40 (quote, 1st paragraph) Katz 1994 Just Do It: The Nike Spirit in the Corporate World, Random House; 41.5 Community Aid Abroad, Australia; 42.2 Simmons Aerofilms Ltd; 43.5 Source: www.Reading.gov.uk; 44.1-9 David Bratley; 46.1 David Bratley; 47.2 Source: Hob-nob Anyone? Reading Football Club Fanzine; 47.3 David Bratley; 47.4 Simmons Aerofilms; 48.2 Trackair; 48.3 David Bratley; 49.4 Courtesy of Newmarket Journal; 50.1, 52.2 David Bratley; 53.3 (top right) Eye Ubiquitous, Corbis; 53.3 (top left) Chris Andrews Publications, Corbis; 53.3 (bottom) Courtesy of Vodafone; 54.2 56.1, 57.4, 57.5, 58.1, 58.2 Bob Digby; 58.3 New Forest District Council; 59.5, 61.3, 61.4 Bob Digby; 62.1 Simmons Aerofilms; 62.2 Bob Digby; 63 (lower left), 63 (lower right) Telegraph Colour Library; 64.1 HSL UK; 64.2 Cartographical Services; 65.3, 66.2, 68.1, 68.2 Bob Digby; 69.3 Telegraph Colour Library; 69.4, 70.2, 71.3, 71.5, 72.1, 72.2, 72.3, 74.2a, 74.2b, 75.3 (top), 75.3 (centre), 76.1 Bob Digby; 76.2 Daintree report, Cairn News, Australia; 77.3 Bob Digby; 78.1 Associated Press; 81.5 Associated Press/NASA/JPL/NIMA; 83.4, 83.5 Associated Press; 84.2 Corbis Images; 85.3, 87.5, 89.2 Associated Press; 90.1, 91.3, 91.4 Ross Parry Picture Agency; 93.4 Pictor International Ltd; 94.2, 94.3, 95.5 Ross Parry Picture Agency; 96.1 Associated Press SABC/APTN ; 96.2, 97.3, 97.4 Associated Press; 99.2 Bob Krist/Corbis Images; 99.3 Associated Press; 100.1 Ross Parry Picture Agency; 100.2 Associated Press; 101.3 Environment Agency; 101.4 Ross Parry Picture Agency; 104.1 Owen Franken, Corbis Images; 104.3 Andrew Laxton, The Mail on Sunday; 114.3 Martin Adler, Panos Pictures; 116.3 (top left) Brian Leng, Corbis Images; 116.3 (top right) Walter Hodges, Corbis Images; 116.3 (bottom left) Ed Kashi, Corbis Images; 116.3 (bottom right) Chinch Gryniewicz, Corbis Images; 124.2 OXFAM; 126a Associated Press; 126b Bojan Brecelj, Corbis Images; 126c Adam Woolfitt, Corbis Images; 131.5 Mark J. Tweedie, Ecoscene, Corbis Images; 133.5 Sean Sprague, Panos Pictures; 142.1 Jane Entwistle; 143 Nick Hawkes, Ecoscene, Corbis Images; 147.3 (article and map) Times Newspaper Limited, 16 February 2001; 147.3 Kevin R. Morris, Bohemian Nomad Picturemakers, Corbis Images; 152.2 (left) Daniel O'Leary, Panos Pictures; 152.2 (right) Jean-Léo Dugast, Panos Pictures; 157.4 Carole Goddard; 158.2 McDonald's; 159.3, 159.4 Carole Goddard; 160.1, 160.2, 161.4 Pictor International Ltd; 161.5 Dave Bartruff, Corbis Images; 163.2 Pictor International Ltd; 166.1 Richard T. Nowitz, Corbis Images; 167 Associated Press; 168.1 Reuters NewMedia Inc., Corbis; 171.2 Marcus Rose, Panos Pictures; 172.1 Charles O'Rear, Corbis Images; 172.2 J.P. Rupp, Black Star, Colorific; 173.3 AFP Photo; 174 (top), 174 (bottom) Paul Glendell, Still Pictures; 175.2 (top) Pictor International; 175.2 (bottom) Dylan Garcia, Still Pictures; 176.1 Don Hebert, Telegraph Colour Library; 177.3 Richard Price, Telegraph Colour Library; 179.4 Mark Edwards, Still Pictures; 180.1 Dominic Harcourt-Webster, Panos Pictures; 197.2 (top left) Nigel Dickinson, Still Pictures; 197.2 (top centre) Mark Edwards, Still Pictures; 197.2 (top right) Mark Edwards, Still Pictures; 197.2 (bottom left) David Drain, Still Pictures; 197.2 (bottom right) Boonsiri-UNEP, Still Pictures; 200.2 Hulton Archive; 202.2 Gareth Davies, NFU; 205.3 George McCarthy, Wild Images, RSPCA Photolibrary; 205.4 Cambridgeshire Libraries; 206.2 (bottom right) Gareth Davies, NFU; 208.3 Panos Pictures; 212.1 Andrew Testa, Panos Pictures; 213.2 Jim Holmes, Panos Pictures; 213.3 Syndication International; 214.1 Jeremy Hartley, Panos Pictures; 216.3, 219.3 Giacomo Pirozzi, Panos Pictures; 220.2, 221.3, 221.4 Jeremy Hartley, Panos Pictures; 222.1 Hulton-Deutsch Collection, Corbis; 223.2 Jaswinder Bhath; 224.1 Eric Whitehead, Cumbria Picture Library; 224.2 Bob Digby; 225.3 National Railway Museum, Science & Society Picture Library; 225.4 London Aerial Photo Library, Corbis; 225.5 Angelo Hornak, Corbis; 226.1 Bob Digby; 226.2 Martin Jones, Ecoscene, Corbis Images; 226.3 Britain on View; 228.1 Bob Digby; 228.3 Britain on View; 229.4 Julie Fryer, Cumbria Picture Library; 230.1, 230.2 Bob Digby; 232.2 Britain on View; 233.4, 234.1 Eric Whitehead, Cumbria Picture Library; 235.3, 235.4 Bob Digby; 237.3 (left) Pictor International; 237.3 (centre) Martin Jones, Corbis Images; 237.3 (right) Bob Digby; 238.2 Cumbria Picture Library; 240.1 Bob Digby; 240.2 Julie Fryer, Cumbria Picture Library; 241.3, 241.4 Eric Whitehead, Cumbria Picture Library; 242.1 Robert Gill, Papilio, Corbis Images; 243.4 Ann Purcell, The Purcell Team, Corbis Images; 244.1, 244.2 Peter Johnson, Corbis Images; 245.4 Jamie Harron, Papilio, Corbis Images.

The illustrations are by Marcus Askwith, Sheila Betts, Maggie Brand, Jeff Edwards, Richard Morris and Tony Wilkins.

The Ordnance Survey map extracts on pages 44-45, 48, 54, 200, and 231 are reproduced with the permission of the Controller of Her Majesty's Stationery Office © Crown Copyright.

Every effort has been made to contact copyright holders of material reproduced in this book. Any omissions will be rectified in subsequent printings if notice is given to the publisher.

The publisher and authors would like to thank the many people and organizations who have helped them with their research. In particular the Lake District National Park Authority, the National Farmers Union, Professor J A Allan (School of Oriental and African Studies, University of London), and Adam Read (consultant, International Waste Management Team, Environmental Resources Management, Oxford).

YOUR COURSE AND THIS BOOK

If you're doing the Edexcel GCSE Specification B, this is the book for you.

The table below provides a summary of the specification, and gives you the appropriate chapters in this book. Each unit in the specification is divided into a number of 'enquiry questions', and these address the main issues involved. These are issues the world faces today, and they are shaping the future.

Core units – *You will study all of these*

Unit A1: Providing for population change

Population dynamics
1.1 How is population changing?
1.2 Why is the population changing?
1.3 What are the social and economic implications of population change?

See **Population dynamics**, pages 102-1125

Population and resources
1.4 What are resources?
1.5 What are the resource implications of population change?
1.6 How are energy resources being used?

See **Population and resources**, pages 126-149

Unit A2: Planning for change

Settlement
2.1 Where shall we build new homes?
2.2 How is rapid growth affecting cities in LEDCs?
2.3 Can urban areas be made more sustainable?

See **Settlement**, pages 6-29

Employment
2.4 How is the global workplace changing?
2.5 What is the impact of new job opportunities in MEDCs?

See **Employment**, pages 30-53

Unit A3: Coping with environmental change

Coasts
3.1 How do physical processes help to create coastal management concerns?
3.2 How do human activities help to create coastal management concerns?

See **Coasts**, pages 54-77

Hazards
3.3 How can tectonic movements create hazards?
3.4 What are the risks associated with flooding?

See **Hazards**, pages 78-101

Option units, addressing the theme 'The use and abuse of the environment'
– *You will study one of the B option units and one of the C option units*

Unit B4: Water

4.1 What issues affect the supply of fresh water?
4.2 What happens when people try to improve their water supply?
4.3 How sustainable is our use of water?

See **Water**, pages 150-173

Unit B5: Weather and climate

5.1 How can weather and climate be a resource?
5.2 How can people modify the weather?
5.3 How can climate change on a global scale?

See **Weather and climate**, pages 174-197

Unit C6: Farming

6.1 What is the impact of modern farming methods?
6.2 What alternative farming methods could be used?
6.3 How can environments be damaged by farming mismanagement?

See **Farming**, pages 198-221

Unit C7: Recreation and tourism

7.1 Why is the countryside being increasingly used for recreation?
7.2 What are the opportunities and challenges that visitors bring to the countryside?
7.3 What can be done to manage the countryside sustainably?

See **Recreation and tourism**, pages 222-245

Contents

SETTLEMENT

Study 1 Where shall we build new homes?
1.1 What is the issue? *6*
1.2 Hands off Ashford *8*
1.3 Developing greenfield sites *10*
1.4 Developing brownfield sites *12*

Study 2 How is rapid growth affecting LEDC cities?
2.1 Why are cities in LEDCs growing so fast? *14*
2.2 How has São Paulo grown? *16*
2.3 What are the effects of rapid urban growth?
2.4 Improving quality of life in squatter settlements *20*

Study 3 Can urban areas be made more sustainable?
3.1 What will our cities be like in the future? *22*
3.2 Why is transport a key issue? *24*
3.3 So how is Boston, USA, solving its urban traffic problems? *26*
3.4 Managing London's waste *28*

EMPLOYMENT

Study 1 The changing world of work
1.1 What job do you do? *30*
1.2 How employment varies around the world *32*
1.3 Global Ford *34*
1.4 The Josh Club – Ford moves into the Indian subcontinent *36*
1.5 Nike – Just Do It! *38*
1.6 Nike – Do It Justice! *40*

Study 2 Reading
2.1 Where are the new jobs? *42*
2.2 Views of Reading *44*
2.3 Up the Royals! – From Elm Park to the Madejski Stadium… *46*
2.4 Getting out of the town centre *48*
2.5 The Oracle Centre – Reading town centre fights back *50*
2.6 The M4 Corridor *52*

COASTS

Study 1 Christchurch Bay
1.1 How can we manage Christchurch Bay? *54*
1.2 Why is there a problem of erosion at Barton-on-Sea? *56*
1.3 How can Christchurch Bay's cliffs be protected? *58*
1.4 Solving one problem, creating another? *60*
1.5 Should Hurst Castle spit be protected? *62*
1.6 Is the cost of protection worth it? *64*

Study 2 The Daintree World Heritage Coast
2.1 Another day in paradise? *66*
2.2 What are the pressures on the Daintree? *68*
2.3 Are we destroying what we most want to see? *70*
2.4 It's not just for the tourists *72*
2.5 Which way now? – 1 *74*
2.6 Which way now? – 2 *76*

HAZARDS

Study 1 Earthquakes, volcanoes and plates
1.1 Where, and why, do earthquakes and volcanic eruptions happen? *78*
1.2 What happens where plates meet? *80*
1.3 Montserrat, a major volcanic eruption *82*
1.4 Can we predict volcanic eruptions? *84*
1.5 Turkey, August 1999, a major earthquake *86*
1.6 Can we predict earthquakes? *88*

Study 2 Flooding
2.1 What causes flooding? *90*
2.2 Flooding in North Yorkshire, 1999 *92*
2.3 What were the effects of the flood? *94*
2.4 Flooding in Mozambique, 2000 *96*
2.5 What were the causes of the flood? *98*
2.6 Can flooding be controlled? *100*

POPULATION DYNAMICS

Study 1 How is population changing?
1.1 The world just keeps on growing *102*
1.2 China *104*
1.3 Where in the world is population growing? *106*

Study 2 Why is population changing?
2.1 Why do some countries' populations grow faster than others? *108*
2.2 What else causes different patterns of population growth? *110*
2.3 Population change in Malawi *112*
2.4 How does migration affect Malawi's population? *114*
2.5 Population change in Germany *116*

Study 3 What are the implications of population change?
3.1 LEDCs – coping with growing numbers of young people *118*
3.2 Developments in Malawi *120*
3.3 MEDCs – coping with an ageing population *122*
3.4 Valuing our elderly population *124*

CONTENTS

POPULATION AND RESOURCES

Study 1 What are resources?
1.1 What are the differences between different types of resources? *126*
1.2 Using resources *128*

Study 2 How are energy resources being used?
2.1 France, an MEDC *130*
2.2 India, an LEDC *132*
2.3 How sustainable are these energy supplies? *134*
2.4 Are biogas plants better than thermal power stations? *136*

Study 3 Will there be enough resources?
3.1 How do we use our resources? *138*
3.2 What does the future look like? *140*
3.3 The way ahead – 1 *142*
3.4 The way ahead – 2 *144*
3.5 Looking back at China *146*
3.6 Population and resources – whose responsibility is it? *148*

WATER

Study 1 Why so much fuss about water?
1.1 What is our main source of fresh water? *150*
1.2 Why do water supplies vary? *152*
1.3 Is there enough water for everybody? *154*

Study 2 Demand is rising. What can we do?
2.1 Providing food for a growing population *156*
2.2 How are urban areas in LEDCs coping with water demand? *158*
2.3 What's happening in MEDCs? *160*
2.4 How has Egypt increased its water supply?
2.5 The High Aswan Dam – do the benefits outweigh the disadvantages? *164*
2.6 Can we use our water more efficiently? *166*

Study 3 Is our use of water sustainable?
3.1 How do we spoil our water? *168*
3.2 Water use – social and environmental disaster *170*
3.3 Global warning *172*

WEATHER AND CLIMATE

Study 1 How can weather and climate be a resource?
1.1 Rain, sunshine, wind or snow *174*
1.2 Attracting tourists – summer sun and winter fun *176*
1.3 Weather, climate and energy – rolling back for a new future *178*
1.4 How do weather and climate influence farming? *180*

Study 2 Should people modify weather and climate?
2.1 Control at the farm gate? *182*
2.2 Urban hotspots – an unintended outcome? *184*
2.3 Acid rain – a spreading menace *186*
2.4 Acid rain in Europe *188*

Study 3 Global climate change
3.1 Global warming – the planet's hottest problem *190*
3.2 Global warming – should we worry about its effects? *192*
3.3 Global warming and the UK – are we winners or losers? *194*
3.4 The crisis in the atmosphere – it's your decision *196*

FARMING

Study 1 The impact of modern farming methods
1.1 Foot and mouth – the last straw? *198*
1.2 How are farming methods changing? *200*
1.3 Farming as an industry *202*
1.4 How does farming change the environment? *204*

Study 2 Alternative methods of farming
2.1 Organic farming – so what's the difference? *206*
2.2 Organic farming – what are the issues? *208*
2.3 Genetically modified foods – the case for... *210*
2.4 Genetically modified foods – the case against... *212*

Study 3 Environmental damage and farming
3.1 Struggling with the land – life in the Sahel *214*
3.2 Farming and drought in the Sahel *216*
3.3 Giving the Sahel a future *218*
3.4 Improving life for the Sahel's poorest *220*

RECREATION AND TOURISM

Study 1 Using the countryside for recreation
1.1 Changing holidays *222*
1.2 To the hills! *224*
1.3 Protecting the countryside *226*
1.4 The Lake District landscape *228*
1.5 A place for everyone? *230*

Study 2 The impact of visitors on the countryside
2.1 What are the benefits of tourism? *232*
2.2 What problems does tourism cause? *234*
2.3 How should tourism develop in the future? *236*

Study 3 Managing the countryside sustainably
3.1 Speed cameras on Windermere? *238*
3.2 Can National Parks be protected? *240*
3.3 Tourism in Zimbabwe – 1 *242*
3.4 Tourism in Zimbabwe – 2 *244*

OS map key *246*   Glossary *247*   Index *253*

SETTLEMENT

Study 1
Where shall we build new homes?

So it won't be long before you leave school and start looking for a job? Depending on job availability you may still live at home. But if not, where do you want to live? In a city? On the edge of a city? Perhaps a village? Will you move into new housing? Are you aware how the building of new homes on a vast scale will impact on the sustainability of both settlements and the countryside? Some villages will lose their rural quality as large new estates are built, others will lose their identity as they are engulfed by urban growth. Consider where you want to live, and then where everyone else wants to live. The government has said that the UK will need over 4 million new homes by the time you are 30.

1.1 What is the issue?

Why do we need so many new homes?

The huge increase in the number of new homes needed is *not* a reflection of population increase as **Figure 2** shows. The reasons for the increase are largely due to:

- More and more young people want a home of their own

- The increasing rate of family break up – it takes more houses to provide for a broken family unit

- An increase in one-parent families

- People living to a greater age and wanting their independence – consider your grandparents and how likely they are to want to live independently in the future.

- Sometimes, immigration into the UK places further demands on housing, though it is more than balanced by the numbers of people who emigrate overseas.

26 November 1996

Government announces need for 4.4 million new homes

IN ITS Green Paper 'Household Growth: where shall we live?' the government stated that 4.4 million new homes would be needed in England between 1991 and 2016.

A target had been set to place 60 per cent of this growth in urban areas, using land that had already been developed but which was no longer used (**brownfield sites**). Some groups have seen this figure as unattainable. The Town and Country Planning Association (TCPA) advocates that cramming as many households into unsuitable urban sites is not sensible. They believe that we should be selective in our choice of urban sites and build in well-chosen edge of town locations, and also establish new settlements along **transport corridors**. The Council for the Preservation of Rural England (CPRE) feel that our countryside is being turned into a landscape of concrete as more and more **greenfield sites** (land not previously developed) are built on. In future housing plans should favour **urban renewal** and **affordable homes**.

The TCPA pointed out that finding suitable sites would be a challenge. It would be especially difficult in southern England and the Midlands, which are already highly built over.

Figure 1 Reporting the issue

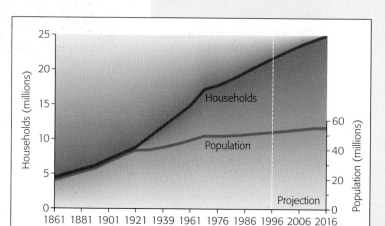

Figure 2 Population and households are growing at different rates

Study 1 Where shall we build new homes?

Did you know?

Extra homes put pressure on the countryside.

- The average house uses up to 60 tonnes of aggregates.
- 4.4 million homes would need 75–90 new quarries to supply raw materials.
- Households with one person (single occupancy) use proportionately more water than households with two or more people (multiple occupancy) and some areas would not be able to meet the increased demands.
- The amount of waste generated will also increase.

Figure 3 Development puts pressure on rural areas

Figure 4 The impact of new housing upon rural settlements

■ How will building so many homes affect you?

It is not just a matter of finding space for new homes. Building on such a large scale has serious consequences for the environment whether these homes are located in urban or rural areas. **Figure 3** notes some social, economic and environmental impacts of development on rural areas. **Figure 4** shows a village where new development has taken place. Environmental awareness, and the sustainability of our cities, are important concepts to consider for the future. How we plan today will affect the environment in 2016, the environment in which we will all live.

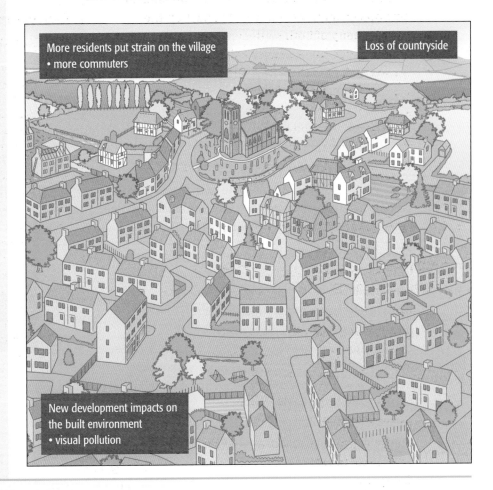

Over to you

1. **a** Do you think that we should accept government predictions for housing demand? Explain your answer.
   b How might patterns in social behaviour change by the time you are 30? What would this mean for the number of houses required?
   c What other factors could alter the trend in housing demand shown in **Figure 2**?

2. Do you feel that we should provide whatever houses people want, or should there be a local or government body responsible for managing housing? Explain your reasons.

3. In your parents' time, people were less concerned about housing developments and the effects they would have. For each of the labels on **Figure 4** add 2 or 3 sentences of explanation thinking about how:
   a loss of countryside affects livelihoods and different interest groups
   b more residents will impact on the quality of village life, local services and facilities, and the environment
   c the new development will change the built environment.

7

SETTLEMENT

1.2 Hands off Ashford

What is it like to live in Ashford now?

Ashford is a town of about 55 000 people in east Kent, a largely rural county. It used to be an important railway town but with closure of the railway works and a decline in rail traffic, Ashford became a depressed town at about the time you were born. But in the 1990s transport developments provided an opportunity for Ashford itself to develop. The M20 reached Ashford. The Channel Tunnel was completed, soon to be followed by the opening of Ashford International Passenger Station. Now less than 2 hours from Paris and 1 hour from London, Ashford has re-invented itself. New housing has been expanding steadily to keep pace with job growth, as companies move to take advantage of the excellent transport there. Unemployment is only 2.4%, very low for Kent as a whole.

As a result, more and more people want to live there. The Borough Council wants growth to be 'sustainable' socially, economically and environmentally. What does this mean?

'Sustainable' development means, basically, one which will last and will not undermine people's quality of life. For instance, an area of new housing may use **brownfield** sites (former industrial land), or it may use a **greenfield** site (an area of farmland on which no building has taken place). Using land that has been used for building before is regarded as more 'sustainable' because it will not use up farmland. In the same way, a housing development that places a primary school in the centre is preferable to one where most children would have to travel by car.

In the UK, housing is regarded as a social right. It has to go somewhere. Yet it arouses all kinds of feeling in people, who may object to any new development in their own area. This has become known as '**NIMBY**'ism – which stands for, 'not in my back yard'. It means that developers all over the UK are likely to meet opposition of some kind when they want to build new housing.

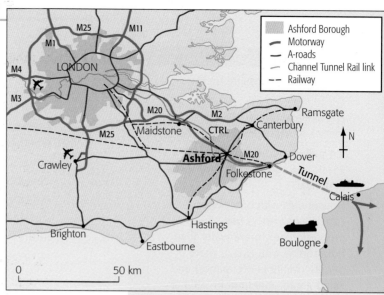

Figure 1 The location of Ashford

Figure 2 Aerial photograph of Ashford showing greenfield development

14th October 1999

SHOCK PLANS FOR ASHFORD
Population to treble

THE CROW REPORT commissioned by John Prescott has targeted Ashford as one of three growth points in south-east England. The recommendation of a threefold increase in population would require a building rate of 2500 homes a year as opposed to the current rate of 700. By 2016, Ashford's population would have increased from 55 000 to about 150 000.

The report also suggested that a Development Corporation was needed to achieve the necessary pace of growth. Such legislation was used to establish new towns such as Milton Keynes.

The borough council is concerned about the lack of brownfield land, only one square kilometre in the borough. It would mean development of greenfield sites on a large scale threatening attractive countryside. It estimated that £2 billion would be needed for infrastructure to support the population increase. The Council is especially concerned about an adequate water supply. Water resources in Kent are already under pressure.

Figure 3 How people see the issues reported

Study 1 Where shall we build new homes?

How might Ashford change?

Like many places, Ashford is changing fast. **Figure 3** is a newspaper article about proposed change in Ashford. Read it carefully, and consider how the town might change in future.

Within a week of the Crow report mentioned in **Figure 3** being published local action groups and residents from the surrounding villages met to decide on a course of action. The council sent a document to the government speaking out strongly against the Crow proposals.

What are Ashford Borough Council's plans?

Ashford Borough Council has its own plans for new developments, shown in **Figure 4**.

Figure 4 Ashford – existing areas and the proposed developments

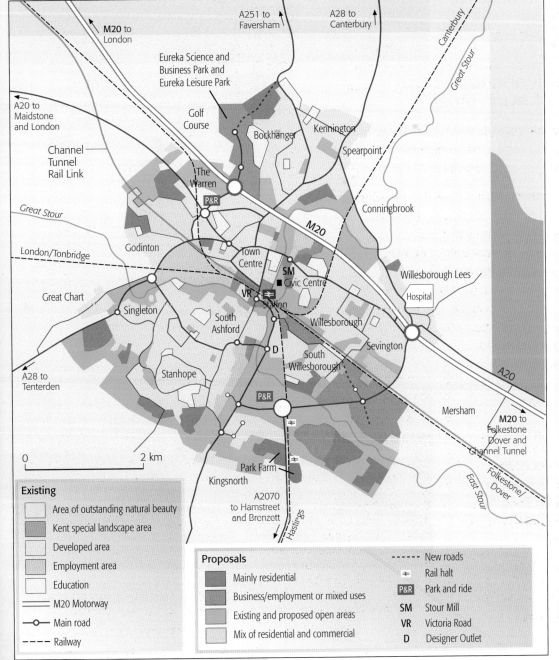

Over to you

1. Find Ashford in your atlas. Draw a sketch map to show its location, and label it with all the advantages you can see for companies wishing to develop there.

2. **a** **Figure 4** shows that most new growth in Ashford is on the south side of the town. Apart from housing, what else is the land being developed for?
   b How far do the proposals fit the concept of 'sustainable development'?

3. Suggest how the new transport links could generate jobs locally and stimulate economic growth. Where else might people find jobs?

4. **a** Do you think that the reaction of the Council and residents to the scale of development recommended by the report is NIMBYism or do you feel it is justified?
   b Would you feel happy about buying a house on the edge of Ashford? Give reasons for your decision.

5. Write an introduction (three or four sentences) for a sales brochure advertising a new housing development in Ashford. Remember to include the advantages of Ashford's location.

SETTLEMENT

1.3 Developing greenfield sites

■ Park Farm – a greenfield site

More land has to be found to meet the demand for new housing. In Ashford, both greenfield and brownfield sites are being used.

Park Farm is one of the greenfield developments begun in the 1990s. It is on the south side of Ashford where most new development has been planned. By the year 2000 Park Farm had 1180 houses and 440 more were under construction. A further 760 are planned for the extension. A primary school opened in September 2000 to relieve pressure on other local schools.

■ What are the advantages and disadvantages of such sites?

Building on such a greenfield site on the urban fringe has several advantages.

There is space for lower density land use. There is also room for community facilities such as shops, a local hall, open space, sports and leisure areas.

The quality of life can be good.

Greenfield sites are cheaper to develop than brownfield sites.

But there are drawbacks.

Distance from the town centre and other important locations increases the need to travel by car. A network of cycle routes, footpaths and improvements to public transport are being explored to help people move round the town.

New main roads are needed to accommodate new residents who also have 1–2 cars per family. The new Southern Orbital road uses up large amounts of land.

Open space between Ashford and surrounding villages is being reduced, which in turn affects the villages.

Figure 1 Variations in design, building material, density and layout in Park Farm help to create a high-quality environment

Figure 2 This large Tesco store serves not only Park Farm but also other new developments in South Ashford

Study 1 Where shall we build new homes?

■ Why is Ashford extending mainly southwards?

Since the 1970s Ashford has been steadily encroaching on the countryside. Look at **Figure 4** on page 9 and see how much land disappears under housing. Some developments such as Singleton and Godinton, started before you were born, are still growing. Only Cheeseman's Green and Lodge Wood await development. Most of the expansion has taken place on the flat, low-lying land on the south side of the town. Not surprising, as the land here is lower grade farmland than north of the town. Nor is it scenically attractive. But close to the northern edge of the town, in sharp contrast, is the chalk ridge of the North Downs, an area of outstanding natural beauty where no major development is permitted. It is an important area for recreation and leisure.

Figure 3 Footpaths and cycle routes are important for access, and help to reduce dependence on cars. They link into the wider network. Note also the 'soft' landscaping.

Figure 4 Wye Downs north of Ashford

■ Is the development sensible?

The floods of October–November 2000 affected homes in Mersham, to the south-east of Ashford. This gave weight to the Environment Agency's demands to call a halt to new housing on flood plains. Doubts were cast on government plans to make Ashford the focus of large scale housing growth in Kent.

Over to you

1. Look at **Figures 1** and **2** on page 10, and **Figure 4** on page 9. Summarize the effects of new development on the outskirts of Ashford. Use a bullet point list.

2. Consider the three advantages and three disadvantages given in the text (page 10) of new housing on the outskirts of Ashford.
   a. Who would probably agree with the list of advantages of living there?
   b. How far do you agree with these?
   c. How do they fit with the housing development nearest to your own home?

3. Look at **Figure 4** on page 9. Explain why villagers in Great Chart (west of Ashford) and Mersham (east of Ashford) might be concerned about the expansion of Ashford.

4. Summarize in a table the advantages and drawbacks of developing greenfield sites.

SETTLEMENT

1.4 Developing brownfield sites

As already mentioned on page 8 not all development has to take up farmland. Throughout the UK, there are large areas of land that have been used for industry or housing in the past, and which are now derelict. These form ideal sites for development of housing and usually result in fewer objections by local people. Such sites are known as brownfield sites. Redeveloping brownfield sites not only reduces the loss of countryside to housing, but is seen as one way in which local planners can make towns and cities better places to live and work. Since the mid-1980s it has become obvious that suitable sites for housing are not as plentiful as was once thought.

Ashford has few brownfield sites. A look at some of these helps explain why the redevelopment of such sites for housing is not always possible. Nor is it sensible. One further problem in the area is that housing is expensive compared to the rest of the UK. More and more younger people cannot afford to live there – yet they are the very people that industries need to work there! As a result, many local councils are concerned about providing affordable housing, especially for young or low-income couples.

Three developments on brownfield sites are shown here – a residential development, a designer outlet and a hotel and conference centre. Consider how these may bring advantages or disadvantages to Ashford.

Brownfield sites

- Contaminated sites (e.g. gasworks, railway sidings) can be too expensive to clean up and prepare for development.
- Site access may be poor.
- Sites may not be in an area where people would want to live.
- The EU has ruled that the government subsidy to developers is unfair.

Figure 1 Problems facing brownfield sites used for housing

■ Stour Mill – a residential development

In December 1998 work started on the redevelopment of the industrial site formerly owned by Rank Hovis McDougall. It is now a residential area with sixty affordable homes for renting. The rest of the development comprises short terraces of two and three storey houses built by a private developer.

Some of these houses front the East Stour River and overlook part of the riverside 'green corridor' which provides an attractive open space in an otherwise built-up area. Paths and cycle routes link it to the town centre and train station which are close by, so you don't need a car either!

Figure 2 Stour Mill – a riverside development. Good design and site layout make high density housing possible

Study 1 Where shall we build new homes?

Figure 3 Designer outlet... with 5 million visitors per year

Figure 4 Plans for new development opposite the International Passenger Station

■ Designer outlet

The site of the disused Kimberley railway works that closed down in the 1980s has now been replaced by a **designer outlet** – that is, a shopping centre where designer label goods are available at cheaper prices. As such, it has become a major visitor attraction. It provides 1000 jobs and brings money into the town. It is also well placed for access from the M20 and close to the international and domestic stations. It should draw 5 million visitors a year. This number of people should stimulate further development in the town centre, only 8 minutes away by shuttle bus.

■ Hotel and conference centre

The prime site opposite the International Passenger Station, and just within the town centre, will not be just for housing. The focus of the area will be a hotel and conference centre and next to it will be a leisure and retail complex, with offices above. Plans also include some riverside housing.

Over to you

1. There are nine other brownfield sites in the town centre that will be variously developed for business, retail, leisure, restaurants, car parking and residential uses. How do brownfield developments in Ashford compare with those in your own area? Give three examples close to you and draw a sketch map to show their location.

2. The biggest growth area of the Kent economy is leisure. Do you think the developments in Ashford will help the economy? Justify your answer.

3. Nationally, would you encourage more or less brownfield development? Why?

4. a Look back over pages 6–13. Write a paragraph which begins – 'Features that lead to the sustainability of towns include ...' and summarize what you think makes some kinds of development sustainable.

   b Now repeat **a** for unsustainable development.

   c How far would you say Ashford was 'sustainable' in the way that it is developing?

SETTLEMENT

Study 2
How is rapid growth affecting LEDC cities?

Towns and cities in LEDCs are growing rapidly, and an increasing proportion of people are living in cities – a process known as **urbanization**. LEDCs are undergoing a period of urban growth and urbanization similar to that in Europe in the nineteenth century.

2.1 Why are cities in LEDCs growing so fast?

In part, the rapid growth of LEDC cities can be linked to the continuing high level of the natural increase of LEDC populations. The natural increase of the population, or the annual **growth rate**, is the difference between the **birth rate** (the average number of births per 1000 people) and the **death rate** (the average number of deaths per 1000 people). Most increase is taking place in cities.

A high birth rate

Although falling, birth rates in many LEDCs continue to be high. The main reason for this is limited birth control, or lack of family planning. Other reasons also contribute to the high birth rate.

Large families are (against all our perceptions) usually much more prosperous, with many hands available to work and share tasks such as water collection. The more children there are to work on the land, the greater a family can cope with harvest.

In spite of huge improvements in attending mothers during birth, and in infant care, **infant mortality** remains high; if many are likely to die in infancy, more births will compensate for loss.

Some religions promote large families.

Figures 1 and **2** Contrasting rural lifestyle and perceived urban lifestyle in LEDCs

A falling death rate

In the past, the very high death rates experienced by most LEDCs tended to counterbalance the effects of the high birth rate. However, today, death rates are falling quite sharply as medical care, sanitation and clean water supply steadily improve. The combination of high birth rates and falling death rates has tended to produce a 'population explosion' in many LEDCs. The natural population increase is, on average, three times higher for LEDCs than for MEDCs. In fact, birth rates in some MEDCs are so low that population is actually falling.

Study 2 How is rapid growth affecting LEDC cities?

■ Migration to the cities

The high rate of natural increase is having an effect on the population growth of LEDC cities but it is also having an effect on the rural areas. So why are cities growing proportionally faster than rural areas? It is largely due to **migration** to the cities. People migrate due to 'push and pull factors':

Push factors – the things that are seen as 'wrong' with the countryside, which encourage some to leave in order to seek better conditions.

Pull factors – the things that people imagine will be better in the cities and which will enable their lifestyle to improve.

Figures 3 and **4** Push and pull factors

Push factors	Pull factors
♦ the poor chance of getting any kind of regular employment or job	♦ better paid factory work rather than poorly paid farm work
♦ the difficulty of getting sufficient land for any kind of successful farming – plots are often very small and tending to get smaller as land is divided between sons on the death of the father	♦ better and more varied food supplies – it is the urban areas that experience food shortages last
♦ relatively few people are able to own their own land	♦ better housing with electricity, piped water, sewerage etc.
♦ near starvation as crops fail due to the weather or as a result of very poor yields	♦ better medical care and the chance of regular and better schooling
♦ increasing soil erosion resulting from over-cropping or over-grazing	♦ the entertainments and more exiting lifestyle of the city
♦ lack of basic services – medical, educational, water, sanitation, transport etc.	
♦ the effects of natural disasters such as drought or flooding	

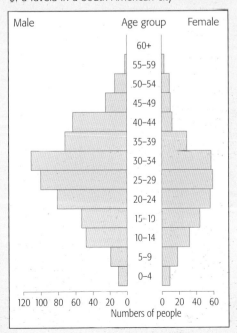

Figure 5 Typical age–sex population pyramid of a favela in a South American city

Over to you

1 a What is meant by 'natural increase of population'?

b Why is there still a large natural increase in the population of many LEDCs even though birth rates are gradually declining?

2 a With a partner, make a list of the reasons why people living in rural areas may no longer want to continue to live there.

b Devise a list of reasons that encourage people to stay in rural areas.

c Which is the longer list, and how does it help to explain why some people stay at home and others go to the city?

d Classify all the reasons in both lists into social, economic and environmental. Which is the longest list?

3 Consider **Figure 5**. Describe:

a which people have moved to the city

b why this is the case

c the potential effects upon rural areas if this trend is typical.

4 a Summarize the 'attractions' of the city for those living in the rural areas. How realistic are these attractions likely to be?

b Classify all the 'attractions' into social, economic and environmental. Which is the longest list?

Settlement

2.2 How has São Paulo grown?

The world's cities are changing. While some cities in MEDCs are growing slowly, others in LEDCs are growing very rapidly indeed. This study looks at São Paulo, one of the world's largest cities, located in Brazil. Figure 1 shows its location within the south-eastern part of the country. Most of Brazil's cities are located on or close to the coast; São Paulo's nearest large neighbour is Rio De Janeiro (**Figure 1**).

Figure 1 The location of São Paulo

Increasing population

In 1950, when your grandparents were children, São Paulo was a relatively small city. Its population was just over 2.3 million. In 2000, this had multiplied nearly 8 times to 17.3 million. By 2015, it is likely to be nearly 21 million. As this study shows, most of the increase is due to the influx of **migrants** from rural areas of Brazil. The maps and photos on these pages show some of the effects of this growth.

Increasing size

Figure 2 shows how far the city has spread as more people have moved there. You can measure how far the city extends from north–south and from east–west. It has become one of the largest urban areas in the world. As it has done so, it has engulfed other settlements around it, which were once small villages or towns in their own right. Now these are just suburbs within the São Paulo

Figure 2 The geographical spread of São Paulo

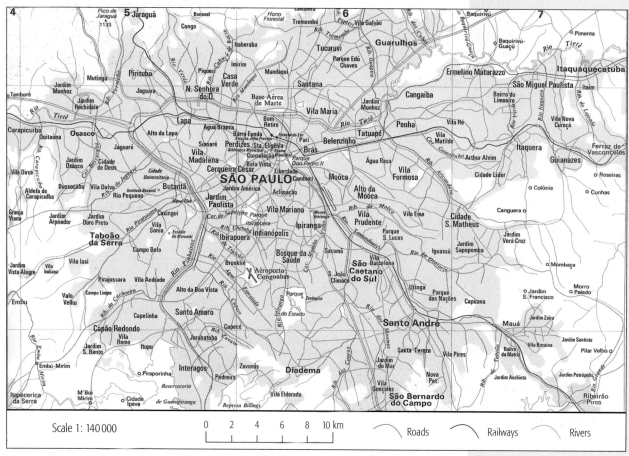

Study 2 How is rapid growth affecting LEDC cities?

urban area. It has therefore become a **conurbation** – that is, a city which has grown so much that it engulfs and absorbs other towns to form a continuous urban area. There is more than one conurbation in this part of Brazil – if you look at an atlas map of Brazil you should be able to spot others.

The increasing size of the city has resulted from the huge concentration of economic growth there. **Figure 3** shows part of the CBD (**Central Business District**) of São Paulo. Almost all of Brazil's major companies have their headquarters here, even though it is not the capital city. You should be able to pick out roughly in which part of São Paulo the photo in **Figure 3** was taken.

Figure 3 The Central Business District of São Paulo

Over to you

1. Using **Figure 1** and your atlas, describe the location of São Paulo **a** globally, **b** within South America, **c** within Brazil, **d** within the particular region of Brazil of which it is a part.

2. Why should most of Brazil's major cities be located on or close to the coast? You should be able to think of several reasons.

3. Study **Figure 2**.
   a Measure the extent of São Paulo i) from north to south, ii) from east to west.
   b How does São Paulo compare in size with other cities that you know within the UK and elsewhere?

4. São Paulo has grown so much that it now forms a conurbation.
   a Using **Figure 2**, name other towns and cities that once surrounded São Paulo, but which have become engulfed by it.
   b Show by a diagram how the city has grown and engulfed other settlements around it.

5. Study **Figure 3**.
   a Identify on **Figure 2** where the CBD of São Paulo is located, and describe how you recognized it.
   b Describe the main features of the CBD as shown in the photo and in **Figure 2**.
   c Explain why many major international companies would be attracted to São Paulo as an attractive economic location.

6. What problems can you suggest could happen in the CBD if São Paulo continues to grow at its present rate?

SETTLEMENT

2.3 What are the effects of rapid urban growth?

São Paulo – a mecca for migrants

São Paulo in Brazil already has a population of over 17 million. Each year, more migrants enter the city so, with the additional increase from births each year among the existing population, the city continues to grow.

Why do people continue to migrate to this city, and what issues does migration bring? All the pull factors and many of the push factors from pages 14–15 are likely to play a part. However, São Paulo is Brazil's greatest industrial city and so the range of job opportunities is greater here than anywhere else. The effects of this rapid growth upon São Paulo are numerous.

- air pollution from industries and traffic
- a severe shortage of land and very high land prices, especially anywhere that is close to the city centre; only the very wealthy can afford to live close to the centre
- a severe lack of open space within the city
- a severe housing shortage, particularly at the lower end of the price range which has led to the widespread development of shantytowns on the outskirts of the city (known locally as favelas)

Figure 1 Some impacts of rapid growth upon São Paulo

What are favelas (spontaneous settlements) like?

As the city has grown, so the pressure on the poorest has been greatest. The search for housing and an inability among the poorest to pay high rents has resulted in an explosion of temporary housing, known as **favelas**. Constructed of any material that is cheap and available, these squatter settlements tend to be of three main types, which reflect how long the community has been established.

1 Very basic shacks built of any recyclable material that new arrivals can find – wood, cardboard, sacking etc. Shelters are erected on pieces of wasteland at the very edge of the city or sometimes on odd pieces of wasteland close to factories. Normally the land is 'waste' because it is of little economic value – perhaps it is badly polluted, or on very steep hillsides or on land that is liable to flooding. These shacks often consist of only one room for living, sleeping and eating, with no electricity, water or sanitation. Any roads are just dirt tracks. City authorities may provide standpipes for communal water supply close at hand. Nonetheless, the lack of sewerage systems results in breeding grounds for disease, and the buildings themselves present a serious fire risk. Residents have no direct access to organized medical facilities, schooling or other services.

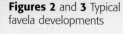

Figures 2 and **3** Typical favela developments

Study 2 How is rapid growth affecting LEDC cities?

Figures 4 and **5** Informal employment

2 Older favelas show evidence of improvement. As incomes become more secure among families, walls are built where once was sacking or cardboard. Individual houses are still likely to lack electricity, water and sanitation, though water taps and 'hook-ups' to electric cables can be achieved in return for fees paid to local councils or companies. Roads are unpaved and become subject to erosion during tropical downpours. Schools become established and some medical care is available.

3 In time, permanent buildings become the norm. Often consisting of several rooms, with bathroom and toilet, permanent kitchens, separate bedrooms, they resemble housing which is poorer perhaps than western European standards, but which has safe amenities and is part of a permanent community. Bus routes pass close by and there are schools and other medical facilities.

Unemployment in LEDC cities

Another major problem associated with the rapid growth of LEDC cities is high unemployment. Although São Paulo is Brazil's largest industrial city, growth in population far outstrips the number of permanent jobs (**formal employment**). Most of those living in favelas are officially 'unemployed' in terms of formal employment. In fact, they are employed but, in the **informal sector** which is huge, jobs such as shoe cleaning, street selling or the small-scale manufacture of items from waste material are available. One major advantage of this huge surplus of labour is that the city is never short of people willing to take on menial or 'dirty' jobs that provide a regular wage.

Over to you

1 Consider the effects of rapid growth on São Paulo shown in **Figure 1**:
   a rank them in order of importance
   b justify your rank order.

2 Consider how the four effects in **Figure 1** might be linked together. Draw a diagram to show how they might be linked, and write a paragraph to explain your diagram.

3 Compare housing and employment in São Paulo against life in rural areas. In what ways do:
   a cities like São Paulo offer improvements over rural areas?
   b rural areas offer advantages over cities like São Paulo?

4 Is life in cities better than rural areas? Use evidence from this case study to support your view.

SETTLEMENT

2.4 Improving quality of life in squatter settlements

Quality of life in LEDC cities

There are two extremes of **quality of life** in São Paulo and other LEDC cities.

At the top, the quality of life is high. Those in highly paid jobs live in city-centre apartments, protected by security guards, with their own recreational facilities, or in large detached houses with security gates, gardens and swimming pools. Again they are protected by security guards and are close to good shops, hospitals and schools.

At the bottom, the quality of life is poor. Remote from the facilities of the city, the poorest have no access to services, no permanent employment and have to rely on the absolute minimum for survival.

In the middle are those with permanent jobs, however poorly paid they are. These are longer-established families, living in the more permanent favelas described on page 19. Their homes have electricity, running water and sanitation, their children attend school, and medical facilities are available locally.

Figures 1 and **2**
Contrasting quality of life in the city

The rapid growth of LEDC cities like São Paulo creates massive problems for the city authorities, trying to provide basic services for settlements around the edge, as well as improving older areas of the inner city. The provision of schooling, medical services, refuse collection and disposal, transport infrastructure etc. never seems to keep up with the continuous inflow of people. The expansion of industry also cannot keep pace with the demand for more and more jobs. For many the quality of life gets worse rather than better.

In the ideal world, LEDC city authorities would like to see the total removal of all shanty settlements or favelas. However, this is clearly not possible. The cost would be huge, and also, no matter how hard the authorities try to solve the problem, the continual arrival of yet more migrants means that they can never catch up.

Study 2 How is rapid growth affecting LEDC cities?

Improving life for the poor

In São Paulo two major schemes have been undertaken by the city authorities.

a) Low-cost improvements to existing housing

Using cheap but permanent building materials such as breeze block, temporary structures are replaced with sturdy permanent houses. Each is connected to electricity and water supplies and proper sewerage facilities. Although the local authority actually builds the houses, finishing off is left to the tenants who pay a low rent for the house. This means that only those with some kind of permanent job can take advantage of this improved housing. Those in the informal sector of employment tend to be by-passed by the scheme.

b) Self-help schemes

Here the eventual cost of the housing is made even lower by encouraging the local residents to undertake most of the improvement work. The local authority simply supply the materials. For example, the local authority provide groundwork needed for the drainage and sewage pipes and the residents themselves undertake the construction of houses. They may even surface the roads. The reduction in costs for local authorities means that limited money can be spread further, thus enabling more homes to be built and more areas improved. Again, a low rent is payable for the finished house, so that again many in the informal sector are prevented from taking advantage of such schemes.

The ongoing problem for all LEDC cities is how to cope with the many thousands of people who continue to live in the worst shanty town conditions and have no kind of permanent job and therefore cannot afford even the lowest rent for any kind of basic improved housing.

Figure 3 Low-cost improved housing in São Paulo with simple concrete structures and basic services only

Figure 4 City self-help schemes under construction in São Paulo; the new houses in the background contrast with the older housing

Over to you

1. Consider the solutions given on these pages of low-cost improvements, and self-help housing. What are their advantages and disadvantages?

2. What else might be done to improve conditions found in favelas? How realistic are these for LEDCs?

3. Consider the following:
   a. raise taxes on the rich to pay for improvements in housing for the poor
   b. increase policing massively to prevent further squatter settlement on spare land
   c. improve investment in rural areas to try to stop rural–urban migration.

Make a copy of the table below to show the advantages and disadvantages of **a–c**, and what might encourage each to happen.

Scheme	Advantages	Disadvantages	What might encourage it to happen
Raise taxes on the rich to pay for improvements in housing for the poor			
Increase policing massively to prevent further squatter settlement on spare land			
Improve investment in rural areas to try to stop rural–urban migration			

4. Which of the three above do you prefer? Explain your answer.

SETTLEMENT

Study 3
Can urban areas be made more sustainable?

Consider **Figures 1** to **5** showing photographs of cities today and think about whether they represent good or poor images of cities; think about whether this is how you like them to be. Cities can be the best of places; they have much to offer. But they can also be the worst of places. They are dynamic places but the pace at which they are growing and changing poses problems that affect different citizens in different ways. In the UK much thought has been given to how we should develop our cities so that their growth is sustainable. Some cities in LEDCs are also thinking along similar lines. This section is all about cities as we imagine them... and how they ought to be.

Figure 1 Living in the city can be a pleasant experience, Savannah, USA

3.1 What will our cities be like in the future?

What are our visions for the future?

What should the world be like in the future? Different people have different ideas about how cities should look in the future. Over time different people have painted their views of an ideal future; in 1516, Sir Thomas More wrote of his vision for an ideal place, known as 'Utopia'. Below is a list of things to consider about how cities might be in future. Think about whether they represent your ideal 'Utopia' by considering them against the five photos.

1. **Quality of life**. Cities are for people. They must provide their citizens with an attractive environment that makes for less stressful day to day living. There should be no social exclusion – that is, no groups should be excluded economically or socially.

2. **Housing**. This should be designed with people in mind. There should be green space around in order to maintain a quality of environment. Access to housing should be for both able-bodied and people with disabilities. The market price of housing to buy or rent should not be beyond people's financial means. There should be no crime, and people should feel confident and free to walk about singly or in groups, day or night.

3. **Social amenities**. There should be access to quality health care and education for all, and amenities within easy reach so that car use is minimized. Reducing car use makes the environment safer for children and older people. Amenities should be accessible for all so that they bring about a community 'spirit'.

Figure 2 The cultural mix – architectural variety in Singapore

Figure 3 There's lots to do... and centres of excellence, Sydney Opera House

Study 3 Can urban areas be made more sustainable?

Figure 4 Traffic – an unavoidable urban problem? London

4 **Employment opportunities**. There should be jobs for all, ideally within the local area so that car use is minimized and people do not have to spend between 2 and 3 hours a day **commuting** – something experienced by many commuters between Ashford and London, for instance.

5 **Environmental quality**. Sustainability also involves concern for resources, the environment and our health. Energy-efficient homes, recycling domestic waste water (and all other urban waste water), even recycling old buildings and using land efficiently – all these are ways in which sustainability can be improved.

6 **Economic cost**. Maintaining and upgrading amenities costs money. How do cities finance projects? In part, this happens by attracting investment for major projects that may have a spin-off effect, bringing other growth with them. It helps too to have a wide range of economic activities and so not be affected by a sudden change in one sector.

7 **Good communication**. Both within the city and externally.

8 **Action**. Visions must be translated into action! Cities need people. People participating, communities working together and having a say in what they would like their town or neighbour to be like. Good management is essential. Without good leaders and planners, nothing will be achieved. Plans must be sound and consider the next generation of urban dwellers.

Figure 5 High density housing, grafitti… signs of energy or blight? Lyon, France

Over to you

1 a With a partner think of a city that you like. Write down the features of your city.

  b Classify your list into things that are linked to people (social), things that are to do with money or employment (economic) and things that are to do with the quality of surroundings (environment). Copy the table below and complete it.

Social	Economic	Environmental

  c Which list is longest, and why?

  d In a different colour pen, write down things that you do not like about cities. Again, classify these in a table.

  e Which list is longest and why?

2 Now look at **Figures 1–5**. Which photos fit the things you like about cities, and the things you do not like?

3 Compare the environment shown in **Figure 5** with that in **Figure 1**. Suggest ways in which quality of life will differ and why well-designed neighbourhoods are essential for sustainability.

4 Economic growth creates job opportunities, which also promote sustainability. Either explain in words or by means of a diagram how this can be so.

5 a Draw up a list of features and qualities in a city that you would ideally like.

  b Where possible, give examples of these in places that you know.

  c Design a plan for your ideal city, showing where things would be – which would be in or near the centre and why? Which would be on the outskirts and why? How would people get from one part of the city to another?

  d Write about 'My Utopia – a city of the future' in 400 words; include your plan of your ideal city.

SETTLEMENT

3.2 Why is transport a key issue?

Transport is essential to make connections for social and economic reasons. Most of us use it daily and all of us rely on it to deliver the necessities of our consumer lifestyle. Why has it become a key issue?

The growth of traffic

Since your parents were children, cities have grown at an alarming rate. In LEDCs **urbanization** will continue apace for some time. City growth has two aspects: the sheer number of people and the vast area occupied by cities. New residential areas on greenfield sites have distanced people from their city centre with its shops, services and places of work. It is often much too far to walk in and people's working hours do not always correspond with bus or train schedules (and these services may not even exist), so people use cars to commute to work. On the **urban fringe** new places of work (such as industrial sites), retail outlets, and leisure centres have emerged. As well as city-bound traffic, there is also now an outward flow (towards the outskirts) of traffic and a lateral one (from outer suburb to outer suburb). **Figure 1** shows the Metro Centre in Gateshead on the outskirts of Tyne and Wear. It attracts people from other suburbs, and also from the centre of Newcastle.

Increasing wealth among the top 60% of wage earners in Britain has led to multiple car ownership per household. More and more families make more trips by car, and fewer on foot, by bike or by public transport. **Figure 2** shows the effects of these changes on the morning commute into Birmingham.

What are the problems?

An increase in the volume of traffic and numbers of car journeys causes a number of problems. Increased pollution (both noise and air), rush-hour congestion, and accidents, especially in winter. People's journeys have been not only lengthened in time but also in distance. And further frustration awaits us at our destination. Where do we park?

Figure 1 The Metro Centre in Gateshead – an out of town shopping and leisure centre

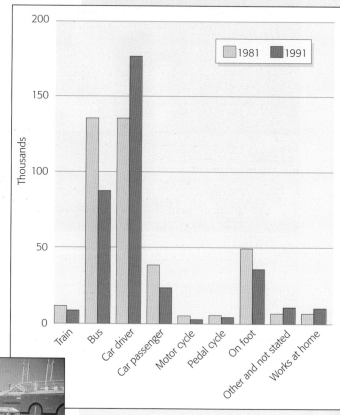

Figure 2 Changes in mode of travel to work of Birmingham residents: 1981–1991

Figure 3 Getting into the city is just the start of the problem, then you have to park

Study 3 Can urban areas be made more sustainable?

Can we make changes?

Traffic congestion is a key issue in all cities and the main culprit is undoubtedly the car. The problem has been brought about by the rapid growth in car ownership and our dependence on the car. In the UK we clock up a higher annual mileage per person in our cars than in any other industrialized EU country. Although many car journeys are essential, uncontrolled growth of car use will deny us the very accessibility to the town or city which the car at present provides, for work and for leisure for example.

Changes in our urban transport system may be difficult to accept but are nevertheless essential. To slow down the growth in car-based travel we need:

- urban transport systems which are sustainable, and accessible to all
- public transport, especially the bus, which must be seen to be an attractive alternative to the car
- walking- and cycling-friendly routes for short distances.

Figure 4 shows the reduction in the share of car transport (rather than absolute numbers of cars) that Birmingham hopes to achieve over 20 years.

But sustainable transport systems involve other changes as well. These include:

- reducing the number of heavy goods vehicles (HGVs)
- efficient highway management
- zero-pollution modes of transport, for example trams, liquid petroleum gas (LPG) buses, battery-powered cars
- land-use planning which either reduces the need to travel (e.g. by providing facilities in neighbourhoods), or permits access by public transport e.g. location of offices near transport interchanges.

Datafile - Birmingham: Public transport

Birmingham City Council has drawn up a transport strategy to tackle the issue of sustainable transport over the next 20 years.

- A large fleet of buses. The Showcase Project introduced in 2000 and now extended to cover 10 routes, has led to an increase in passenger numbers: 27% on Route 67.
- 11 key rail routes with frequent services for commuters. The Jewellery Line came into operation in 1995. Rail passengers have increased since then.
- 48% of all trips into the city during peak commuting hours are made by bus and rail.
- In May 1999 a light railway, the Metro Link between Birmingham and Wolverhampton, has provided a service (20.4 km in 35 minutes). To be extended by 2005.
- Taxis: 950 licensed hackney carriages; 3700 private hire vehicles.

Purpose	City centre			Outer area		
	Current	Target		Current	Target	
		Existing	New Development		Existing	New Development
Work	50%	45%	40%	70%	65%	60%
Shopping	25%	25%	25%	50%	45%	40%
Leisure	25%	25%	25%	70%	70%	70%
Other	30%	25%	20%	45%	40%	35%

Figure 4 Twenty-year targets for car use set by Birmingham City Council; some percentages could decrease, but not all

Over to you

1 a How do:
   i) you get to school?
   ii) your parents get to work?
   b Are there alternative means for each of you? Are they affordable? Are they practicable? Why?

2 a Sustainable transport policies seek to increase the number of footpaths and cycleways for local journeys:
   i) how well does your local area provide cycleways?
   ii) do many people use them? Why?
   b What are your views on increasing the use of cycle lanes in British cities?

3 Using **Figure 2** describe the changes in transport modes used by Birmingham residents to get to work over the decade 1981–1991.

4 Brainstorm your class for ideas:
   a to discourage people from using their cars to go into towns and cities
   b to make public transport an attractive alternative mode of transport to car use

5 Traffic hold-ups are a social, economic and environmental issue. Draw a spider diagram to show this.

3.3 So how is Boston, USA, solving its urban traffic problems?

Solving urban traffic problems is no easy matter. Consider the space, the disruption, not to mention the cost of introducing a new transport scheme. Providing more road space is not necessarily the solution, it will simply generate more traffic! So what are the alternatives? Boston (USA) has come up with some innovative ideas!

Boston's problem

Boston, Massachusetts, is the regional capital of New England in north-eastern USA. Its population grew from 2.5 million in 1960 to over 5 million by 2000, and growth has brought with it a major traffic problem on the Central Artery which takes highway 1–93 through downtown Boston. 1–93 is a six-lane elevated highway built in 1959, but it can no longer cope with the volume of traffic using it. The datafile shows the scale of the problem.

Datafile – the Central Artery

- traffic increase:
  1959 75 000 cars per day,
  1989 190 000 cars per day
- by 1989, 8–10 hours per day of traffic jams were causing financial losses to business
- the bridge needed repairs; would this be money wasted?
- by 2010 it was estimated there would be stop-go traffic for 14–15 hours a day
- further financial losses would be incurred if Logan International airport did not have an improved link with the city
- $500m would be lost in late deliveries, accidents and wasted fuel.

Figure 1 The original Central Artery

Boston's solution

The city authorities decided to handle a future increase in traffic demand by demand management, which means carrying more people rather than by carrying more cars. An integrated plan was drawn up to increase ways of travelling within the city and beyond. The key element of the plan was the construction of a new Central Artery, but this time underground. It was made financially possible by a Federal Highways Bill, which allocated money to road construction. The elevated highway would be left in place until work was completed on the new artery to cause minimum disruption.

Study 3 Can urban areas be made more sustainable?

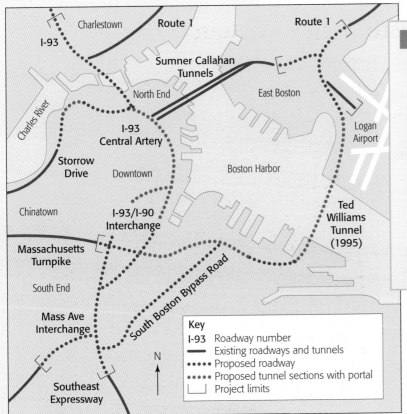

Figure 2 The Central Artery and Tunnel project

Details of the scheme

- 8–10 lane highway
- will carry 245 000 vehicles a day
- morning and evening rush hours each only 2–3 hours
- traffic will move at 30 mph
- video cameras, electronic message boards
- High Occupancy Vehicle lanes
- Reduced local traffic will speed up flow
- Don't need to stop at pay tolls

The new Central Artery and Ted Williams Tunnel will form part of an inter-modal transport scheme – that is, people will use several forms of transport instead of just one:

- new harbour ferry services such as the airport water shuttle will take people off the roads
- increased choice of travel mode will reduce commuter traffic – commuter ferries, double-decker trains, minibus, **car pooling** agencies, additional park and ride
- a new fast train service was started in November 2000 between Boston and Washington providing improved links to other regions.

How sustainable is Boston's plan?

When completed in 2004, Boston's plan should reduce carbon monoxide pollution by 12%, and remove an eyesore (the elevated highway) which will release 13 ha of land. Redevelopment such as low rise housing and retail developments, a multimillion dollar convention centre and landscaped green areas are planned.

Figure 3 Work on the Ted Williams Tunnel

Over to you

1. Apart from the construction costs of the Boston scheme, what other costs can you think of for users of the system?
2. **a** Express in your own words how demand management should reduce the number of cars on the roads in the future.
   b What might prevent this from happening?
3. 70% of airport traffic in Boston comes from the west and south. Explain how new road improvements should improve the quality of life in Downtown Boston and East Boston.
4. Originally the scheme was considered useless, but is now seen as a means to revitalize Boston. Apart from relieving traffic congestion, how is the scheme improving the sustainability of the city?

SETTLEMENT

3.4 Managing London's waste

What's the problem?

Waste is a mounting problem in London, increasing at the rate of 3–4% each year. It is reaching crisis point. In 2000, Londoners generated almost 17 million tonnes of household waste. Changing lifestyles are much to blame. The advent of the supermarket and self-service stores has meant that most consumer goods are now pre-packaged. The expanding selection of magazines for all tastes generates yet more waste whilst TV adverts tell us we cannot live without their latest product. We must throw away the old and buy the new. Our wheelie bins remind us: we live in a discard society. But the disposal of this waste is an ever-increasing challenge.

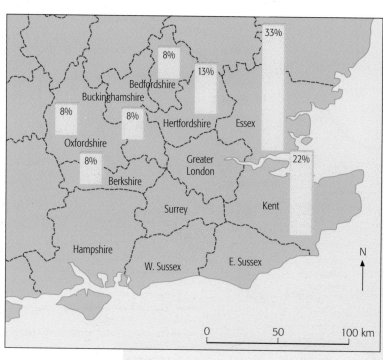

Figure 1 Percentages of London's waste going to some of the Home Counties

How sustainable is London's waste management?

Just where, and how best, to dispose of London's waste had already become a problem by the mid-20th century. **Landfill**, the cheapest means of disposal was seen as the best way. But until very recently, only two sites existed in this densely built-up city. So since the 1960s increasingly longer journeys have been made to reach disposal sites in the Home Counties. Despite the use of the high-density rail network and the river for bulk transport of waste to distant sites, road plays a major role in refuse collection and in other local transfers. Quite clearly, transporting waste is a major environmental issue. And so are landfill sites!

Crisis point: more waste, less space

New laws are about to make important changes. By 2015 the export of waste to landfill sites must be cut to one third of the present amount. And an EU Directive decrees that the amount of biodegradable material at landfill sites must be reduced by 2020.

Figure 2 London and waste management

An enormous task

Infrastructure includes:
- 500 waste collection vehicles, barges, specialist transporters in use every day
- 17 major transfer stations
- 45 civic amenity sites
- 2 incinerators
- 8 major recycling centres
- 2 compost sites
- 3 landfill sites
- 2 energy from waste plants

but more is needed.

The Thames helps – Cory Environmental move 600 000 tonnes of waste by river each year. By road, this would need 400 lorry movements a day!

Study 3 Can urban areas be made more sustainable?

Figure 3 Percentages of current treatment processes in London

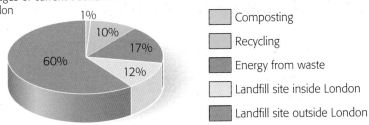

- Composting
- Recycling
- Energy from waste
- Landfill site inside London
- Landfill site outside London

How should we deal with waste in future?

Figure 3 shows what currently happens to London's household waste. However action to protect the environment is clearly forcing change.

- Where possible, recycling must be increased, despite the cost. Individuals can help by buying recycled goods.
- Composting must be increased, e.g. community composting for the neighbourhood and individuals, composting organic waste.
- Waste must be viewed as a resource: recycling and burning for energy have value.
- Money must be invested in new technology.
- Local processing facilities must be built.

No matter what steps are taken London will always be dependent on the Home Counties taking some of its waste. But people involvement (helping to cut down on the amount of waste generated) and making greater use of waste as a resource can improve the social, economic and environmental sustainability of London.

Figure 4 SELCHP (South East London Combined Heat and Power). This waste-burning power station provides electricity for homes in South East London

Figure 5 The changing dustbin: 1900, 1956, 2000

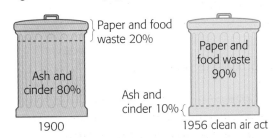
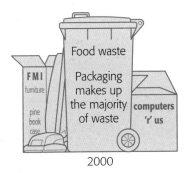

- 1900: Ash and cinder 80%, Paper and food waste 20%
- 1956 clean air act: Paper and food waste 90%, Ash and cinder 10%
- 2000: Food waste, Packaging makes up the majority of waste, computers, furniture, pine bookcase

Over to you

1. Draw a diagram to show the knock-on effects on the environment of transport to landfill sites in the Home Counties. Add the effects that the landfill sites themselves have on the environment.
2. Suggest how waste disposal can have a positive impact on the social (jobs, health) and economic (revenue from waste) sustainability of London.
3. Study **Figure 5**.
   a. Suggest reasons for the changes in waste thrown into bins from 1900 to 2000.
   b. Complete the diagram by adding a drawing to show how you think Londoners may have to sort their waste ready for collection in 2015. Explain your drawing.

EMPLOYMENT

Study 1
The changing world of work

Most people around the world need to work on a daily basis in order to survive. But the world of work is constantly changing. The number of people working in the manufacturing and service-providing sectors has increased in many countries. This shift is partly due to the growth of **transnational corporations**. What are some of the issues associated with these transnational corporations?

1.1 What job do you do?

The types of jobs that people do can vary from a footballer, to an IT consultant, a market trader, or a subsistence farmer. Some people choose to work **full-time**, and others to work **part-time**.

- Full-time work in Britain usually involves people working over 36 hours per week. Many jobs demand that people work for many more hours than this each week. Full-time work is paid either weekly or monthly. Although hours are usually fixed, many people have to work more in busy times, with no extra pay. However, full-time staff usually benefit from paid holidays and paid sick leave.

- Part-time work is when people work for perhaps one or two days a week, or for a few hours every day. You may have a friend who does a 'paper round' or works in the local supermarket on a Saturday. These are examples of part-time work. People usually receive hourly rates of pay rather than salaries, and are paid according to the number of hours they work. Generally part-time workers receive few benefits.

People's jobs can be classified according to whether they are involved in producing goods or services. All types of jobs can be divided into one of four categories as shown in **Figure 1**.

Primary: Jobs based around the collection or harvesting of raw materials – e.g. farmer, miner, and lumberjack.

Secondary: Jobs involved in manufacturing, or making things, e.g. carpenters, builders, and potters.

Tertiary: Jobs involved in providing a service, e.g. teacher, librarian, financial adviser, waiter, or supermarket worker.

Quaternary: Jobs involved in providing information and expertise, e.g. an IT consultant, and market researcher.

Some jobs fall between categories. For example, a sawmill takes felled trees (a primary product) and partly processes them into planks or wood pulp before going to a furniture or paper factory (secondary products). These are known as primary processing industries. You might be able to think of other examples that fall between categories.

Figure 1 Classification of jobs

Figure 2 Advertisements for full- and part-time jobs

Study 1 The changing world of work

Part-time supermarket worker

'I currently have a part-time job working at a branch of Waitrose in Twickenham. I am contracted to work on a Sunday. I arrive at work for a 10am start and finish at about 5pm. My job involves taking stock from the warehouse to the shop-floor and putting as much of it on the shelves as possible. I have an aisle that is mine for the day and I have to make sure that it is kept fully stocked all the time. This can be hard work as we sell a lot of goods during the day and I am constantly filling the gaps!'

Edward Kaviko, farmer, Uganda

'I own 160 cattle which I use for food, milk, and income. The cattle are Friesians and local cross-breeds, which give very good yields of milk. I use one piece of land as a grazing area for the cattle for two weeks, then I fence it off and transfer them to another field to give the land a rest. I spray the cattle every week with tic spray. Normally I sell 40 litres of milk every day to a milk vendor. One litre fetches 500 shillings. If I need to sell cattle I take them to the local market.'

Street trader in Bangkok

'I came to Bangkok in 1993 from my village in northern Thailand. I sell goods near the Royal Orchid Sheraton Hotel where there are lots of business people and tourists. I get a good trade, selling silk goods made in local factories. Some stalls sell fake goods like fake Rolex watches, but I sell proper goods. I sell silk ties for 150 Baht (about £2.50) and people tell me that these are ten times more expensive in London!'

Robert Jennings, electronic engineer

'I usually get in to work for about 8:30am and spend my morning checking through the software programs that my team were writing the day before. We nearly always encounter problems and I spend most of the morning correcting them. After lunch in the canteen we test out the software. We're working on a satellite navigation system so we take the project car out on the test route. After an hour or so we get a good idea of what modifications are needed to the software so we head back to the office and I start coding new software for the project. I'm usually away by about 4:30pm.'

■ Working within the law

When you start working you will probably reply to an advert similar to those in **Figure 2**. If you are successful, you will be asked for your National Insurance number and will pay income tax. Legitimate businesses like these are part of what we call the **formal economy**, where companies or organizations trade legally and are registered with the government.

Many people work in the **informal economy**, operating outside legal restrictions and unknown to (or at least not tracked by) tax or health and safety offices. Examples include a shoe shiner on the streets of São Paulo in Brazil, an illegal street trader on Oxford Street in London, or someone selling cigarettes purchased in France and brought back to Britain. All of these jobs earn money that is not subject to income tax, *as authorities cannot measure or track it!*

Figure 3 The jobs that people do…

Over to you

1. Classify all the jobs mentioned on pages 30 and 31 into formal and informal, using the table below.

	Formal work	Informal work
Full-time job		
Part-time job		

2. Draw a table showing the advantages and disadvantages of formal and informal work, using these examples and others from your friends or family to help you.

3. a Classify the jobs shown in **Figures 1**, **2**, and **3** into primary, secondary, tertiary, or quaternary, using the table below.
   b Add a further three examples of jobs in each category.

Primary	Secondary	Tertiary	Quaternary

4. Carry out a survey of the pupils in your class by asking:
   a what jobs they do
   b what jobs adults at home do
   c where each person works
   d how far they travel to work and how they get there
   e whether they travel as part of their job or not.

   Process the results, classifying them into full-time or part-time, and primary, secondary, tertiary, or quaternary. Use the results to produce a class display of the jobs that people do in your local area.

5. How might your survey produce different results in different parts of the urban or rural area that you live in? Give examples of how and why these might vary.

EMPLOYMENT

1.2 How employment varies around the world

■ United Kingdom – a service-based economy

The UK economy is now mainly a service-based economy. Industries such as banking, insurance, and business services account for the largest proportion of the **Gross Domestic Product** (GDP). The City of London (**Figure 1**) is one of the world's leading financial centres, and employs several thousands of people, as well as controlling the finance and livelihoods of millions of others.

Farming and other primary industries do not now employ many people in the UK. Agriculture has become highly mechanized, producing about 60% of our food needs with only 1% of the workforce. Manufacturing has continued to decline, with closures of car assembly plants and shipyards. Meanwhile, the service economy, including quaternary employment, has grown.

Figure 1 The City in London – where several of the world's largest financial companies are located

Figure 2 Employment change in the UK since 1970

	1970	2000
Primary	4%	2%
Secondary	48%	42%
Tertiary and Quaternary	48%	56%

■ South Korea – a newly industrializing country (NIC)

Since the 1960s South Korea has undergone an 'economic miracle'. It has grown from an isolated, war-torn, and financially dependent country to become the world's eleventh largest trading nation.

In the 1960s the South Korean government planned a series of five-year economic plans. Its development has grown in three stages:

Stage 1 – 1960s: It had few natural resources and little technical expertise, so early development focused on manufacturing, such as textiles, that used large amounts of labour.

Stage 2 – 1970s: Investment grew in heavy industries, such as shipbuilding, iron and steel, and cars. The government planned to export these to the world market.

Stage 3 – 1980s and 1990s: Since the 1980s there has been a move towards hi-tech industries such as computers and electronics.

Today, South Korean companies such as Samsung, Hyundai, and Daewoo are household names, exporting TVs, cars, hi-fis, and computers. Output per person has grown from US $87 in 1963 to US $10 000 in 1995, and South Korea's economy grows by about 8% each year.

Figure 3 Cars made in South Korea, ready for shipment

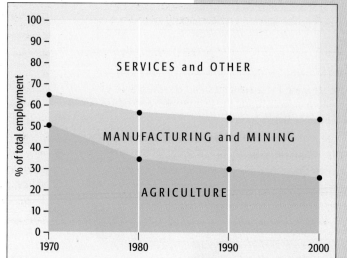

Figure 4 Changes in employment in South Korea 1970–2000

Study 1 The changing world of work

■ Ghana – a struggling economy?

Ghana is a country on the West Coast of Africa with a population of 14 million people. Most people live in rural areas and make a living from the land. The economy is based upon subsistence farming, which accounts for 46% of wealth and employs 60% of the workforce. It revolves around food crops and cocoa production.

Memuma Aberake talks about her working day in Ghana

'By nine o'clock I've many jobs to do – it might be farming or house building. Once every four days I go to the bush to collect firewood. I have to walk about two miles. I bring back a big load on my head – it's very heavy to carry. If I start at nine I won't come back until one. Sometimes I collect guinea corn stalks from my husband's farm for firewood instead.'

In recent years, Ghana has begun to develop rapidly at an average annual rate of nearly 5%, growing by expanding its exports. It still depends heavily upon cocoa, gold, and timber, and most of its growth has yet to spread to other areas of the economy.

Figure 5 Watering the crops in Ghana

Figure 6 Employment in Ghana in 2000

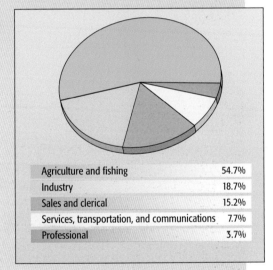

Agriculture and fishing	54.7%
Industry	18.7%
Sales and clerical	15.2%
Services, transportation, and communications	7.7%
Professional	3.7%

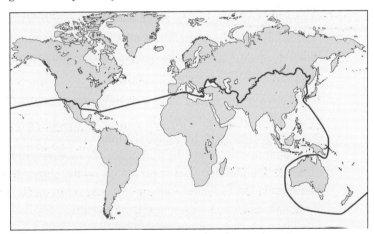

Figure 7 The north-south divide

Over to you

1 The City of London employs *'several thousands of people, as well as controlling the finance and livelihoods of millions of others'.*

  a What jobs do people do in the City? Give several examples.

  b How do these people 'control' the lives of millions of others? Identify as many ways as you can, and suggest how this happens.

2 Using an atlas to help you:

  a locate the UK, Korea, and Ghana on a blank world map

  b label each country with its economic characteristics, such as employment.

3 Make a copy of the following table. On it, label the features of the economy of the UK, Korea, and Ghana.

	UK	Korea	Ghana
Main employment types			
Main exports			
In which employment is growth currently taking place?			

4 In 1980, the **Brandt Report on World Development** drew a line dividing the world into a rich 'north' and a poor 'south' shown in **Figure 7**.

  a Which category would the UK, Korea, and Ghana fit in?

  b Is there still a 'north-south' divide on the basis of these countries? Use other examples to support your answer.

EMPLOYMENT

1.3 Global Ford

The Ford Motor Company was established in a converted wagon shop in Michigan, USA in 1903; it had just 10 employees. Today, Ford has more than 345 000 staff members in over 200 countries, North America, Europe, and Mexico being the major locations for research and assembly. In addition, 10 000 companies worldwide supply Ford with goods and services, from manufacturing car parts to providing food for the staff canteens. The company has expanded to become the world's largest producer of trucks and the second-largest producer of cars.

Strategies for continued success

Study **Figure 1**. Notice anything? In fact there is no odd one out because Ford made all four cars, even though they bear different names. With the end of the twentieth century approaching, Ford realized that sales of their brand were beginning to reach maximum potential. A new strategy was needed to build on the company's incredible success.

Ford's answer was to buy other car companies that were either competition or that allowed Ford access to a new market. The purchase of Jaguar in 1997 gave Ford direct access to the European luxury car market, while the purchase of Land Rover in May 2000 allowed entrance into the lucrative 4 x 4 market. Buying companies and brands like this gives Ford an advantage over its competitors, since names such as Jaguar and Land Rover already have an established position in their sectors of the car market.

Global strategy

The Mondeo, produced in 1993, was Ford's first 'world car'; it was designed for a world market and assembled at locations across the globe (see **Figure 3**). Combining the best designs and technology from Ford's operations in both America and Europe, the Mondeo was the result of a $6 billion research programme. It was widely popular in Europe but never took off in the USA.

Undeterred, Ford has continued in their pursuit of global product strategies. With the launch of the Ford 2000 initiative in 1995 Ford announced they were going to consolidate their global operations and concentrate on developing cars for worldwide markets. Ford predicted this strategy would:

- result in corporate growth
- allow them to produce cars for world markets

Figure 1 Which is the odd one out?

The one millionth Ford Focus from Ford's Saarlouis plant came off the assembly line at the German facility today. This production milestone for the Ford Focus came at a time of record demand for the compact mid-range car. Over 160 000 customers in Europe chose Ford's top-selling model in the first quarter of 2001, a 15% increase over the previous year.

The Ford Focus has won over 50 national and international awards across the world since its launch in 1998. It was the first model to win the 'Car of the Year' title in both Europe (1999) and in the USA and Canada (2000). In addition to the Saarlouis factory, the Ford Focus is also built in Valencia, Spain and at the Ford plants in Wayne, USA and Hermosillo, Mexico. Ford Focus production worldwide has already passed the two million mark.

Figure 2 Worldwide success. Ford press release, April, 2001

Study 1 The changing world of work

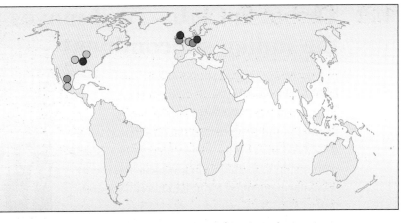

1. Assembly takes place at Genk in Belgium, Kansas City in the USA, and Mexico.

2. The four-cylinder engine is being made at plants in Bridgend in the UK and at Cologne in Germany for Europe, and at Chihuahua in Mexico for North America.

3. The executive car is being made at Cleveland, Ohio, for both the US and European-produced cars.

4. Manual transmissions are being made in Europe at Halewood in Merseyside and at Cologne; gearboxes are made in the USA.

Figure 3 The Ford Mondeo is assembled in a truly global way!

Damian, a brand specialist in Essex.

" I was based at Ford of France's Bordeaux plant and helped launch a system for reducing machinery downtime at the plant…
I was responsible for co-ordinating local implementation and organizing a training programme for 3000 French-speaking shop-floor workers. "

Cam, a personnel officer in Ohio.

" In April 1973, I was hired at the Ford transmission and chassis plant in Sharonville, Ohio. I was placed on the Ford College Graduate Program and started my career as a production supervisor. While assigned to the Sharonville Plant I also held positions in Industrial Engineering and Quality Control before I received an opportunity to be reassigned to Human Resources in January 1976. Since then, I have held positions in Salaried Personnel, Salaried Administration, Employee Relations and Labor Relations. "

Figure 4 Not everyone working for Ford makes cars and trucks!

- cut development costs as a result of knowledge-sharing
- allow them to allocate resources wherever they were needed to best serve market needs
- ultimately allow them to establish a dominant global presence.

Ford considered this global strategy to be the best way they could maximize sales and beat their competitors. Have their efforts paid off? In 2000, Ford achieved a record volume of sales, they had both the world's best-selling car, the Ford Focus (see **Figure 2**), and truck, the Ford F-series.

What is a transnational corporation (TNC)?

Transnational corporations (also known as multinational companies) are large companies that operate in more than one country. They may have their headquarters in a MEDC and then have a number of factories and outlets around the world. Many TNCs have utilized cheap labour, especially in countries in South East Asia and other LEDCs, as an alternative to the rising costs of labour in their home countries. They grow by:

a expanding, e.g. Coca Cola

b merging with competitors, e.g. Ford

c buying a wide range of companies. GlaxoSmithKline for instance, who produce a wide range of goods, from Lucozade to pharmaceuticals.

Over to you

1. Write a one-sentence definition for a transnational corporation.
2. a What benefits might a company such as Ford have from being distributed worldwide?
   b What difficulties might arise from transnational distribution?
3. Imagine that you run a company that manufactures clips for securing hubcaps. You employ just 15 people. Your main client is the Ford Motor Company. Your machinery breaks down and you are late with an order to Ford. Ford decides to find another supplier.
   a How might your workers be affected by this decision?
   b How might the Ford workers be affected by this decision?
   c What changes can you suggest that might help you to negotiate a new contract with Ford?
   d How might these changes affect your company?
4. A fall in Ford's profits could result in a decision to make changes to the structure of the company. In pairs discuss the possible changes. Describe:
   a where changes might be made
   b who will make the changes
   c who will be affected by the changes.

EMPLOYMENT

1.4 The Josh Club – Ford moves into the Indian subcontinent

In 1995, the Ford Motor Company established a partnership with Mahindra and Mahindra Limited in India. It was designed to help Ford get a foothold in the huge Indian market. In 1996 the partnership launched its first vehicle, the Escort, specially engineered for Indian conditions and tastes.

Ford have invested heavily in India and in March 1999 they opened a 170 billion rupee (about £2.5 billion) integrated hi-tech manufacturing plant at Maraimalai Nagar, 45 km north of Chennai, employing 22 000 people. The new plant occupies a 250 acre site and covers all stages of car manufacturing. Over 100 000 cars will be produced every year. Ford India has 80 suppliers and employs a large local skilled workforce.

Figure 1 The launch of the Ikon in India

Why should Ford wish to invest in India? In 1999, India's population passed the 1 billion mark. Although many think of India as a largely rural and poor country, it is one of the world's top 10 economies. It has several major industrial areas, and a high-earning middle class. Labour costs are low – approximately 10% of those in the UK – and it provides Ford with a presence in the Asia-Pacific region.

Ford's new tactics

Ford's new factory at Maraimalai Nagar has developed several new practices designed to promote Ford's image in India. It claims to have the following – (think about why these might make a difference to people working there and to Ford).

- A 'flat' organization structure, and one common canteen for all; all employees on first-name basis.
- Women are 20% of workforce (industry average 5%).
- All Ford employees, including shop-floor technicians, are encouraged to make decisions and take responsibility in how the factory is run.
- Only 100 employees are trained overseas.
- An open-learning centre enables anyone working there to take courses within the plant about technical, business, and communication issues, during working hours.

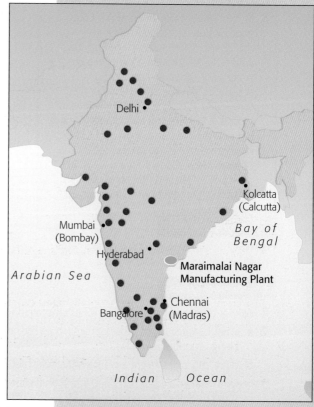

Figure 2 Manufacturing plant and Ford dealerships in India

Study 1 The changing world of work

■ Ford's progress

The first product to be manufactured at the facility is the Ikon (**Figures 1** and **3**), a new car which is of 70% Indian origin. 22 849 were sold in the first year, generating a turnover of £220 million.

Every car has a soul. An identity. That calls, inspires, and thrills you. For the Ford Ikon, it's josh. A feeling that sends excitement coursing through your veins. That makes your blood flow faster. A feeling that makes you look life in the eye and plunge right in. It's hard to explain. But easy to feel.

The Josh Club is a unique circle of lively Ford car owners, bringing together dynamic and spirited go-getters who live life with passion. As a privileged member, you can expect a range of exciting offers on premium products and services through our unique tie-ups. And more exciting events, fun, and adventure activities. Make sure you have registered for your Josh Club ID card!

Figure 3 The Ford Ikon and the sales pitch that goes with it in India

Ford's progress in India has been good; they claim that 'the year 2000 was excellent for us'. The company believes that 2001 will see them dominate the 'Ikon' sector of the Indian car market and move to selling the Indian-built Ikon in other South East Asian countries, such as Bangladesh.

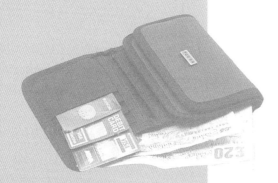

Over to you

1 Why are the following important in Ford's decision to locate in India? Explain as fully as you can –
- A population of over 1 billion
- India is one of the world's top 10 largest economies
- India has a high-earning middle class
- Labour costs in India are only 10% of those in the UK
- Ford wanted a foothold in the Asia-Pacific region.

2 What is a 'flat organization structure'? Why should Ford want to advertise it among its workers?

3 a What image do large companies such as Ford have when they develop factories overseas? Why is this?

  b Why should Ford want to advertise the following features of working conditions in its new factory in India?

- Women are 20% of workforce
- All Ford employees, including shop-floor technicians, are encouraged to make decisions and take responsibility in how the factory is run
- Only 100 employees are trained overseas
- An open learning centre enables anyone working there to take courses within the plant about technical, business, and communication issues, during working hours.

4 Compare Ford's advertising in India (**Figure 3**) with a car sales campaign that you know of in the UK. What is similar about the India campaign? What is different?

EMPLOYMENT

1.5 Nike – Just Do It!

Nike products are instantly recognizable all around the world with their distinctive 'swoosh' tick and the slogan 'Just Do It'. The company dates from 1964 when Phillip Knight (an athlete turned businessman) came up with the idea of importing running shoes from Japan. Japanese shoes were cheaper because labour was, in those days, cheaper in Japan. His aim was to compete with German brands such as Adidas and Puma.

Today, Nike is the largest seller of sports footwear, clothing, and accessories in the world. It operates in over 140 countries around the world employing about 8000 people in the USA alone. Most products are made by independent contractors, who are paid by Nike to make goods for them. Some Nike clothing is made in the USA, but nearly all training shoes are made elsewhere by several hundred thousand workers, mainly in South East Asia.

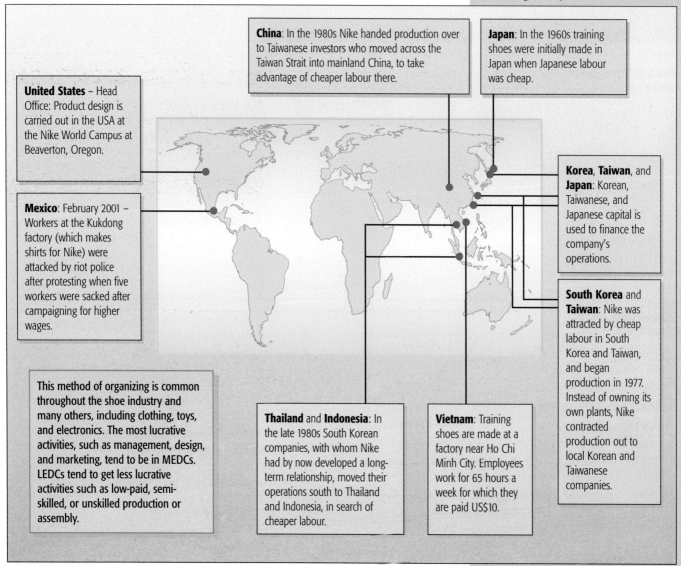

Figure 1 The production of Nike training shoes is a global operation

China: In the 1980s Nike handed production over to Taiwanese investors who moved across the Taiwan Strait into mainland China, to take advantage of cheaper labour there.

Japan: In the 1960s training shoes were initially made in Japan when Japanese labour was cheap.

United States – Head Office: Product design is carried out in the USA at the Nike World Campus at Beaverton, Oregon.

Korea, **Taiwan**, and **Japan**: Korean, Taiwanese, and Japanese capital is used to finance the company's operations.

Mexico: February 2001 – Workers at the Kukdong factory (which makes shirts for Nike) were attacked by riot police after protesting when five workers were sacked after campaigning for higher wages.

South Korea and **Taiwan**: Nike was attracted by cheap labour in South Korea and Taiwan, and began production in 1977. Instead of owning its own plants, Nike contracted production out to local Korean and Taiwanese companies.

This method of organizing is common throughout the shoe industry and many others, including clothing, toys, and electronics. The most lucrative activities, such as management, design, and marketing, tend to be in MEDCs. LEDCs tend to get less lucrative activities such as low-paid, semi-skilled, or unskilled production or assembly.

Thailand and **Indonesia**: In the late 1980s South Korean companies, with whom Nike had by now developed a long-term relationship, moved their operations south to Thailand and Indonesia, in search of cheaper labour.

Vietnam: Training shoes are made at a factory near Ho Chi Minh City. Employees work for 65 hours a week for which they are paid US$10.

Study 1 The changing world of work

Working conditions investigated

Training-shoe factories in Asia have been investigated by a number of organizations who campaign for human rights, all of whom have been shocked by the working conditions of employees. Many workers are paid low wages, as little as US $1 a day in Indonesia, which is still poor even though the cost of living is less there. There have been reports of neglect of health and safety, physical, and sexual abuse of workers and persecution of workers who have tried to organize unions.

Figure 2 Extracts from an interview with a child worker at the Nike factory in Kukdong, Mexico, January 2001.

> I am 15 years old, I work at Kukdong. I make 352 pesos a week (about 50 pence an hour). I work 10 hours a day. I work from 8am to 6pm. I am an operator in Line 5, embroidering sleeves. I work five days a week. I don't make enough money.

> The Korean managers haven't abused me, but I've seen how aggressive they are with other co-workers of mine. They yell at them very aggressively.

> I am an operator, so I spend all the time seated. When there is no work, I get bored. The former supervisor didn't allow me to lean my face against my hands. She wanted me to keep sitting straight all the time. I couldn't even bend down.

> I don't like to have lunch there, because the food is very bad. They only give us rancid food, the meat is bad. Also, the food is tasteless. There are no more than three water fountains (for over 800 workers). Sometimes there is no water in these fountains. If we are thirsty, our mouths get dried-up.

Production labour	$2.75
Materials	$9.00
Rent, equipment	$3.00
Supplier's operating profit	$1.75
Duties	$3.00
Shipping	$0.50
Cost to Nike	**$20.00**
Research and development	$0.25
Promotion and advertising	$4.00
Sales, distribution, admin	$5.00
Nike's operating profit	$6.25
Cost to retailer	**$35.50**
Retailer's rent	$9.00
Personnel	$9.50
Other	$7.00
Retailer's operating profit	$9.00
Cost to consumer	**$70.00**

Figure 3 Where the money goes: a breakdown of a US $70 pair of Nike 'Air Pegasus' training shoes

Over to you

1. Use the information in **Figure 1** to produce a time-line showing Nike's changing production locations.
2. **a** Why should Japan, Taiwan, and Korea have been early locations in which to make Nike trainers? Why are they not used now?
   b Using **Figure 1**, say which countries have replaced them, and why.
3. **a** Using **Figure 3**, draw a pie chart to show the proportions of a $70 pair of Nike trainers that go to i) Nike's producers ii) Nike iii) retailers.
   b Who seems to be making most? Why?
   c At what stage of the process do you think the most monetary value is added to the product?
4. Say whether you think the working conditions at Kukdong are good, reasonable, or bad. Explain why you think this is.

EMPLOYMENT

1.6 Nike – Do It Justice!

Nike has been targeted by many human rights organizations worldwide since it came to light that 'unfair' working practices have been taking place in factories around the world where Nike products are made. **Figure 1** shows a leading newspaper report highlighting working conditions in Nike factories. Nike have made their position clear: *'We don't pay anybody at the factories and we don't set policy within the factories; it is their business to run'*.

■ Campaigning against Nike

A host of organizations have launched campaigns to encourage Nike to improve the conditions for workers in factories making their products. The web site 'Just Do It! Boycott Nike' (as shown in **Figure 2**) is one such campaign. It is supported by 53 members of the US Congress and many civil liberties groups including: Justice Do it Nike; US Campaign for Labour Rights; Global Exchange; Press for Change; and The Clean Clothes Campaign. Each suggests a number of strategies to combat the unfair working practices in 'sweat-shops'.

Nike shoe plant in Vietnam is called unsafe for workers

UNDERMINING Nike's boast that it maintains model working conditions at its factories throughout the world, a prominent accounting firm has found many unsafe conditions at one of the shoe manufacturer's plants in Vietnam.

Figure 1 *The New York Times*, front page article, November 8th 1997

Figure 2 An anti-Nike website campaign

Don't buy NIKE products: Nike is one of the biggest training shoe manufacturers in the world. If we can change Nike's unfair practices, other shoe companies will be forced to improve theirs.

Join A Nike Protest: There are many Nike protests organized in towns and cities across the country.

Send a letter to the Prime Minister: Tell him that we need his help in asking Nike to address these unfair practices in South East Asia.

Send a letter to Nike's Chairman, Mr Knight: Tell Mr Knight that you are unhappy with their practices. Ask Mr Knight to accept independent monitoring, accept a living wage standard, and expand Nike's Code of Conduct. Remember to include a pair of Nikes with your letter. If the shoes are brand new, make sure to ask for your money back.

Name	Purpose	Web page
Just Stop It!	Campaigns to persuade Nike to improve their labour practices.	http://www.caa.org.au/campaigns/nike/
Mike & Nike	Never-seen-before excerpts of Michael Moore's interview with Nike Chairman Phillip Knight. 'I think Nike's CEO Phil Knight regrets he said the things he said when I interviewed him last year.'	http://www.dogeatdogfilms.com/mikenike.html
The Clean Clothes Campaign	The Clean Clothes Campaign is an international network with the goal of improving the working conditions in the garment industry worldwide.	http://www.cleanclothes.org/companies/nike.htm
Nikewages.org	Focuses on the lives of sweat-shop workers in Nike's Indonesian shoe factories.	http://www.nikewages.org
Nikeworkers.org	Offers information and resources on Nike worker conditions.	http://www.nikeworkers.org/index_pop_up.html
Nike Beautiful?	A collection of pictures aimed to break the heart.	http://www.ibiblio.org/stayfree/ads/nike.html
Nuke Radioactivewear	A take on Nike intended to blow you away.	http://www.cybold.com/nuke/index.html

Figure 3 Internet campaigns targeting Nike

Study 1 The changing world of work

Figure 4 Some of the organizations and programmes concerned with workers' rights and opportunities that Nike have been involved with.

- ▲ The Fair Labour Association is made up of consumer, human, and labour rights groups, as well as six leading clothing and footwear manufacturers and retailers, including Nike. It is a non-profit organization established to protect the rights of workers worldwide.
- ▲ The Global Alliance for Workers and Communities (GA) was established to improve the lives and future prospects of workers involved in global production and service supply chains. Although Nike was a founding member, the GA is an independent and objective group.
- ▲ In China, Nike and World Vision have created after-hours, free education programmes for workers in footwear factories.
- ▲ In Pakistan, Nike is a member of a coalition whose aim is to eliminate child labour from the industry.
- ▲ In Thailand, Nike and a local charity have established 'Nike villages'; stitching centres formed in rural areas that act as focal points for employment and other opportunities.
- ▲ Working with the Vietnam Women's Union, Nike has established a scheme to provide money to help rural women start up new businesses.

'Wages may be low, but people want these jobs'

'The overwhelming share of workers in our subcontract factories there (Indonesia) have had a positive experience, as evidenced by the fact that the turnover rate in those factories is the lowest in the business... The workers, if you will, vote with their feet. It is clear to them that manufacturing jobs pay a steady wage, and offer the kinds of benefits that are prized in a country where half the workforce is still earning a subsistence income on a farm, often with neither running water nor electricity'.

Dusty Kidd, Nike Public Relations, in a letter to American Friends Service Committee, Oregon about conditions in Indonesian factories.

■ Nike's response

Nike has responded to these campaigns, as in **Figure 5**. Nike says:

- the wages that it pays are fair and are linked to each country's cost of living
- working practices are far worse in most other companies than in their own factories
- they do not employ children or operate sweat-shop conditions.

Nike also point to their own Code of Conduct, established in 1992 when they found that local factory conditions were often poor. This sets out global standards which factories manufacturing Nike products have to meet. Nike says these standards are often higher than local regulations demand. In addition, Nike has worked with several organizations to improve the work environments and lifestyles of the people who make their products, as shown in **Figure 4**.

What do you believe? What else would you like to know? Would what you know alter your shopping habits? In 2001, a research report showed how three-quarters of people in the UK thought it was wrong that clothes should be produced in such conditions. But less than one-third said that it would affect their buying habits, or force them to change brand.

Figure 5 What Nike says in reply

Over to you

1. Consider the evidence for and against Nike given in these pages. Which evidence seems stronger? Why?
2. Use the information provided to draft a statement for Nike to use at a press conference, answering charges of unfair practice.
3. How far would either Nike's own attitudes, or the campaigns against Nike, affect your buying habits? Why?

EMPLOYMENT

Study 2
Reading

Reading has expanded to become one of the most desirable locations for businesses and one of the top retail centres in the UK. This growth has brought plenty of new job opportunities and Reading now boasts very low unemployment figures. What are the reasons behind Reading's success? Has Reading's expansion caused any problems?

2.1 Where are the new jobs?

Reading is a thriving town with a population of about 145 000 people located in the heart of the Thames Valley close to the M4 motorway in South East England. Heathrow Airport is only 27 miles away and London 41 miles. Reading is a competitive business town that attracts national and transnational companies and is also a major shopping centre.

The economy in Reading was traditionally based around manufacturing industry ('bulbs, beer, and biscuits'), but in more recent years the town has become a base for finance and technology industries. In 1996 there were some 3800 businesses in Reading of which 86% were service-based, employing some 74 000 people.

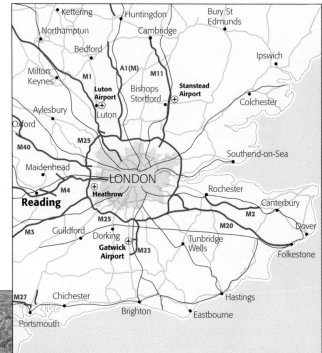

Figure 1 Where is Reading?

Figure 2 Employment by sector in Reading (1998)

Sector	% of workforce
Business services	24.9
Wholesale/retail	16.6
Finance	8.7
Transport and communications	7.8
Education	7.4
Manufacturing	6.9
Public administration	6.8
Health and social work	6.8
Other services	6.4
Hotels/restaurants	4.3
Construction	3.4

Study 2 Reading

Office (town centre)	£22
Office (out of town)	£24–£27
Retail	up to £200
Industrial	£8

Figure 3 Cost per square foot of renting business property in Reading in 1999

Reading is rapidly expanding with a growth of industrial estates, retail parks, and business parks around the edge of the city. There are now five main industrial estates, seven retail parks, and seven business parks (with one under construction). Some of these can be seen in the photographs on pages 44–45. The cost of renting business space in Reading is shown in **Figure 3**.

Reading has attracted a large number of high technology (hi-tech) industries in recent years. However, many of these companies require far fewer workers than the declining 'old' manufacturing industries. **Figure 4** shows the top ten largest employers in Reading. As can be seen, the major employers in Reading are service-based industries with Microsoft being the largest hi-tech employer. Other hi-tech companies currently based in Reading include: Racal Telecom (telecommunications), Energis (telecommunications), and ICL (IT services).

Figure 4 Reading's top ten employers, 2001

Employer	Approximate no. of employees in Reading	Activity
Reading Borough Council	5500	Public administration
Reading University	5000	University
Prudential Assurance	4000	Insurance
Royal Berkshire and Battle Hospital NHS Trust	3500	Health care
Royal Mail	2500	Postal service
Compaq Computers	2000	Electronic equipment
Foster Wheeler	1950	Engineering
British Gas (HQ)	1500	Exploration of gas
Microsoft (HQ)	1100	Computer software
Savacentre	1100	General stores

At the beginning of the twenty-first century, Reading is a large, busy, prosperous town, with more jobs on offer than there are unemployed people. It draws people in from miles around for employment, education, shopping, and entertainment. To the south, the east and the west there are junctions with the M4 motorway. Its railway station has direct trains to most parts of the country, including Gatwick Airport, and it has a coach service to Heathrow Airport. Paddington Station is less than half an hour away. Reading has paid a price for its success – constant rebuilding has swept away many good buildings and reminders of the past. Now, at last, we are taking better care of what does remain.

Figure 5 How Reading Borough Council is looking to the future, while looking after the past

Over to you

1. Describe the location of Reading using **Figure 1**.
2. **a** Using the information in **Figure 2** draw a pie chart to show the different employment sectors in Reading.
   b Work out the percentage of workers in Reading who work in the primary, secondary, tertiary, and quaternary industrial sectors.
3. Using **Figure 3** work out the cost of a 10 000 square foot business premises per year:
   a in Reading town centre
   b on the outskirts of Reading
   c in a retail park on the edge of Reading
   d on an industrial estate in Reading.
4. Traditionally the most expensive sites to rent would have been in the town centre of Reading. Why do think that this is not the case now?
5. Prepare a fax to the Managing Director of COMsoft suggesting possible benefits and disadvantages of locating their European headquarters in Reading.

EMPLOYMENT

2.2 Views of Reading

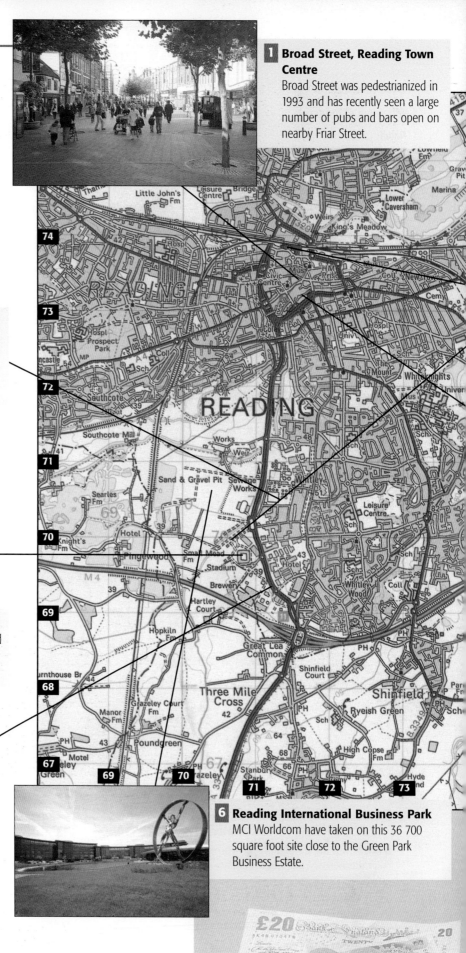

1 Broad Street, Reading Town Centre
Broad Street was pedestrianized in 1993 and has recently seen a large number of pubs and bars open on nearby Friar Street.

9 Brunel Retail Park and Reading Link Retail Park
Offering nearly 200 000 square feet of retail space, many large stores have located here including Safeway, Halfords, Currys, and PC World.

8 Madejski Stadium
The new home of Reading Football Club was opened in September 1998. The site also includes a retail park, conference centre, and luxury hotel.

7 A33 Relief Road
Opened in June 1999 the A33 relief road has improved access to Reading town centre from Junction 11 of the M4 motorway and offers good transport links to a collection of business and retail parks.

6 Reading International Business Park
MCI Worldcom have taken on this 36 700 square foot site close to the Green Park Business Estate.

Study 2 Reading

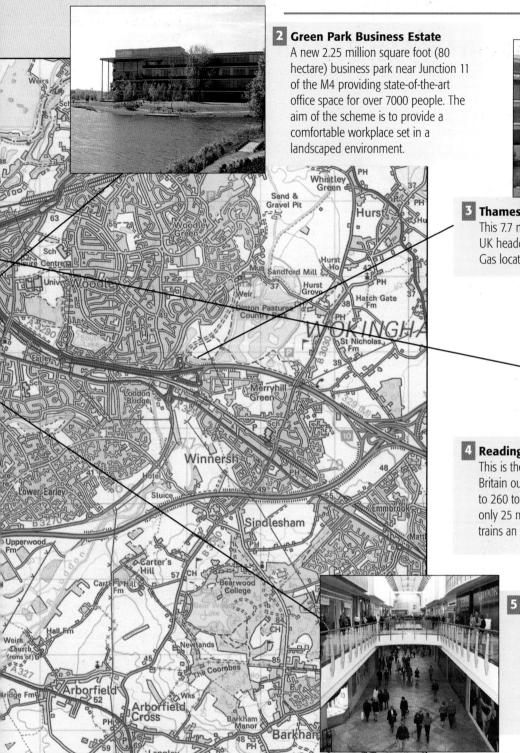

2 Green Park Business Estate
A new 2.25 million square foot (80 hectare) business park near Junction 11 of the M4 providing state-of-the-art office space for over 7000 people. The aim of the scheme is to provide a comfortable workplace set in a landscaped environment.

3 Thames Valley Business Park
This 7.7 million square foot business park has the UK headquarters of Microsoft, Oracle, and British Gas located there.

4 Reading Railway Station
This is the second busiest railway station in Britain outside London and offers direct access to 260 towns and cities. London Paddington is only 25 minutes away and there are up to eight trains an hour at peak times.

5 The Oracle Centre
The new £250 million 700 000 square foot shopping and leisure complex on the River Kennet opened in September 1999. The development includes two major department stores and 90 retail units, as well as a multiplex cinema and riverside restaurants and cafes.

Over to you

1. Match photos 1, 4, 5, and 8 with the map using six figure grid references.
2. What does CBD stand for? Describe the characteristics of the CBD. Compare the CBD of Reading with the CBD in your home town.
3. Identify new development.
   a. Suggest impacts on the CBD
   b. Why do you think that they have caused an impact?
4. Select one of the sites shown as a location for a new ice rink and multi-entertainment centre. Justify your choice of this site.
5. Create an advertising poster encouraging a company to locate on this site.

EMPLOYMENT

2.3 Up the Royals! – from Elm Park to the Madejski Stadium…

■ Goodbye to Elm Park

In 1996 the decision was taken to move Reading Football Club from their home of 102 years at Elm Park, about 2 miles west of Reading town centre. Elm Park was tucked between the A329 Oxford Road and the mainline railway at the back of a large residential area. Throughout the 1990s traffic congestion and restricted parking on match days caused problems for both local residents and the Football Club.

■ Why the need to move?

In the early 1980s Reading FC seemed to be going nowhere. Attendances were falling and the Chairman sold the club to a local property developer who investigated selling the Elm Park site. Plans were drawn up to move to the south of Reading where there was a lot of wasteland close to the motorway. However, as the property market collapsed Reading FC was again sold and the plans to move were dropped.

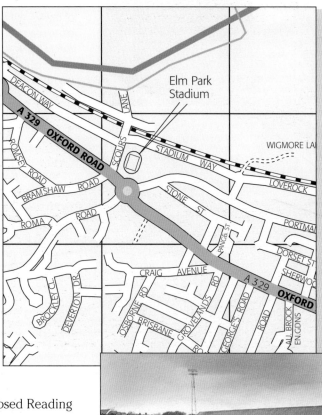

Reading FC was then bought by John Madejski. He did not like the Elm Park site and knew that for the club to ever have a chance of promotion they needed a new stadium. There were several reasons for this.

- Legislation requires football clubs to have all-seater stadia soon after being promoted to Division 1 of the Football League. It would have been difficult and expensive to do this at Elm Park.
- Moving to a large site has allowed the club to build commercial ventures – a conference centre, a 'megastore', bars, and catering facilities.
- The new site was to be shared with other sports teams as tenants – originally, Richmond Rugby Club and currently London Irish Rugby Club.

Figure 1 Elm Park – the old home of the Reading Royals

Southern Reading was still the preferred site for the new stadium and the A33 position was the only realistic site in Reading. The council had been using the stadium site for landfill, but it was now full and needed to be put to an alternative use. The council was therefore supportive of the club's requirements, and sold the site to them for £1 – although it needed to have millions spent on it to clean it up. The club also contributed towards the cost of building the A33 relief road, something the council wanted but which was too expensive for them to build on their own.

Study 2 Reading

Into the Mad House!

Figure 2 What the fans say... from 'Hob-Nob Anyone?', Reading Football Club Fanzine

'A Premier ground for, err... a mid-table division two side!'

The Madejski Stadium (known affectionately as the Mad House by fans!) is the new home of Reading FC. Named after the Reading FC Chairman, John Madejski, the Stadium was built on the edge of Reading on the new A33 relief road close to Junction 11 of the M4 motorway. It is a 25 000 all-seater stadium offering a range of facilities, including the Royal Berkshire Conference Centre for up to 700 people, and a luxury 150-bedroom hotel.

Figure 3 The all-new 25 000-seater Madejski Stadium

Figure 3 The all-new 25 000-seater Madejski Stadium

Over to you

1. Look at **Figure 1** (the location of the old stadium) and **Figure 2**. What problems do you think that the location of the Elm Park stadium could have posed for:
   a Reading FC?   b local people?
2. What advantages would the move have for both the football club and the fans?
3. Local people living in the Tilehurst area of Reading (near the old Elm Park Stadium) made many complaints to Reading FC on match days. Make a list of the problems that residents may have encountered and how these problems would arise.
4. Look at the OS map (pages 44–45) and suggest reasons why the site at Rose Kiln Lane on the A33 (grid reference: 708698) was a suitable location for the new stadium.
5. Are any people disadvantaged by the new location?
6. What opportunities does the new football stadium offer for:
   a Reading FC?   b local people?   c local businesses?
7. Using your atlas measure the distance from London to Reading and work out the distance that London Irish rugby fans would have to travel to see their team now. Then:
   a prepare a letter from a fan to the club expressing your views
   b write a response from the club.

EMPLOYMENT

2.4 Getting out of the town centre

What's at the edge?

Out-of-town shopping has been growing in Reading over the last ten years. Before the completion of the new A33 Relief Road (see map on pages 44–45) the existing A33 into Reading, from the M4 junction 11, was a busy main road that led into the town centre. Vacant land towards the south of the town therefore seemed an ideal location for new out-of-town retail units.

Safeway was one of the first companies to open a large hypermarket on the corner of Rose Kiln Lane and the A33. The close proximity to a large area of suburban housing, Reading University, a large industrial estate south of the town centre, and a busy main route into Reading offered an ideal site for Safeway.

The site at Rose Kiln Lane expanded quickly with several other stores joining Safeway. These included PC World, Currys, and Halfords. The site at Rose Kiln Lane now offers lots of retail outlets as well as several car dealerships. The shops are usually found inside large warehouse-style frameworks, which are well-lit and spacious with wide aisles. Outside are large sprawling car parks and attractive landscaping.

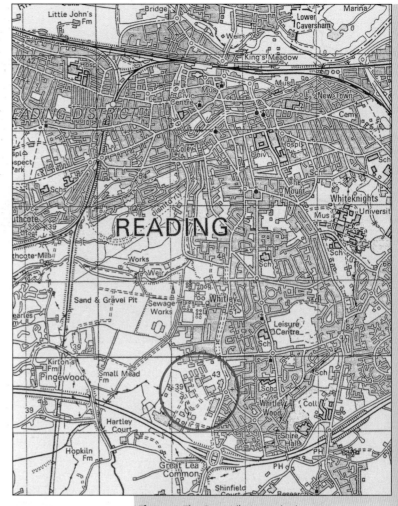

Figure 1 The Rose Kiln Lane site in 1991

Figure 2 Aerial photograph of the Rose Kiln Lane site 2001

Figure 3 The Safeway superstore at Rose Kiln Lane

Study 2 Reading

■ The growth of out-of-town shopping

Figure 4 Shopping on a traditional high street in 1988

Out-of-town shopping became popular in the prosperous 1980s. Many retail companies, especially supermarkets, found that town centre stores were crowded and cramped and did not offer shoppers an 'enjoyable' experience. Part of the problem was due to an increased volume of traffic in town centres, and small, expensive car parks that restricted shoppers.

Supermarkets found that cheaper land on the edge of urban areas coupled with new housing developments offered attractive sites for new stores. Supermarkets could locate near to a large population of potential customers, offer free parking and have larger, more 'welcoming' stores. Ideal locations were often close to motorway junctions or ring roads as these offered good accessibility and the opportunity to attract passing trade.

The emergence of supermarkets on the edge of our towns and cities and their huge success has driven other companies to follow their lead. Marks and Spencer have a programme of opening out-of-town stores, often next to a large supermarket, and other companies, such as Halfords and Currys, are moving away from high street stores altogether.

■ The bid-rent theory

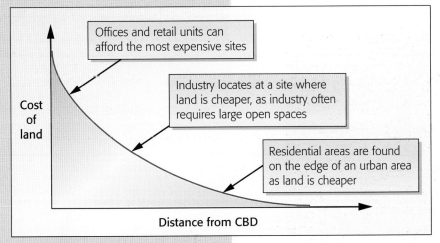

Figure 5 The bid-rent theory

This model is based on the idea that land at the centre of an urban area is the most sought-after. This is because many companies believe that they will attract more customers in the CBD (central business district) where most people either work or go to in order to shop. As you move away from the CBD land becomes cheaper as it is not in such high demand. Supermarkets were among the first retail companies to take advantage of the benefits of the cheaper land available at the edge of urban areas.

Over to you

1. When Safeway opened a store at Rose Kiln Lane they had many things that they wanted. Write a shopping list of things that Safeway would have wanted at a new site.

2. Study the **Figures 1** and **2**. What advantages does the Rose Kiln Lane site offer to retail companies?

3. Draw a sketch of the aerial photograph. Label your sketch with the identifiable features of the retail units at Rose Kiln Lane.

EMPLOYMENT

2.5 The Oracle Centre – Reading town centre fights back

Reading currently ranks 13th in the list of top retail centres in the UK, just below Edinburgh and ahead of Liverpool. This is a rise from 26th place in 1998, and with the Oracle Centre now fully open Reading hopes to rise to a top ten ranking. How can it keep its position with new retail developments opening a short distance away?

Increasing competition from out-of-town retail centres on the fringe of Reading meant the old town centre was under pressure to modernize to attract customers back into the heart of Reading. The out-of-town sites offered good access by car, excellent free car-parking facilities, and large, airy, well-lit shops on modern retail estates. They were safe and free of the pollution and crowds found in many town centres.

Reading has answered with the £250 million Oracle Centre. It spreads over 22 acres of land that previously housed an old NCP car park, a collection of small shops and businesses, and a compound for housing buses at night. This part of Reading had been deteriorating for years and was becoming dangerous at night. This large area of land, just a few minutes walk from Broad Street, with the River Kennet flowing through it, and backing on to large retail stores such as John Lewis (Heelas) and Debenhams, offered the potential for large-scale re-development.

Figure 1 The Oracle Centre, Reading

Figure 2 Shops in Friar Street, Reading, close to the Oracle

- The number of people living within a six-mile radius of Reading town centre is 302 875.
- The number of people living within a ten-mile radius of Reading town centre is 463 639.
- There are three times more high-income families in the catchment area of the Oracle compared with the national average.
- There are twice the number of young professionals with high incomes than the national average.
- 71% of Reading women work, which is 15% above the national average.
- Over 80% of the catchment area are car-owning households, 15% higher than the national average.
- Expenditure on restaurant and café meals is 11% higher than the national average.
- Evening restaurant use is 13% higher than the national average.
- 22% visit a pub on a weekly basis.
- The Oracle captures a lucrative shopping audience thanks to the affluence of Reading and the Oracle's catchment area.

Figure 3 Why was Reading chosen for the Oracle development?

Study 2 Reading

Major stores
Boots
Debenhams
House of Fraser
Audio Visual
Carphone Warehouse
Dixons
Orange
Photo Optix
Vodaphone
Books & Music
Bookworld
HMV
Waterstones
Fashion Accessories
Claire's Accessories
Salisbury's
Sunglass Hut
Food
Millies Cookies
Thorntons

Footwear
Dolcis
Dune
Ecco
Jones Bootmaker
Sole Trader
Gifts and Cards
Bureau
Clinton Cards
Discovery Store
Natural World
The Disney Store
The Gadget Shop
Wax Lyrical
Health & Beauty
Bath & Body Works
Crabtree & Evelyn
Origins
Supercuts
The Perfume Shop
Jewellery
Beaverbrooks
Essential
Goldsmiths
H Samuel
Leslie Davis
Ladies Fashion
Bay Trading
Dorothy Perkins
Etam
InWear
Jeffrey Rogers
Karen Millen
Kookai
Mango
Morgan
New Look
Oasis
Rogers & Rogers

Room London
The Vestry
Top Shop
Wallis
Lifestyle
Blue Tomato
Muji
The Pier
Whittards
Mens Fashion
Burtons
Ciro Citterio
Envy
Matinique
Monserrat
Suits You
Top Man
Mixed Fashion
Eisenegger
French Connection
Gap
H&M
Lacoste Boutique
Levis
Red/Green
Tie Rack
Zara
Opticians
David Clulow
Vision Express
Sports & Outdoor
Allsports
Outdoor Venture
Sports Soccer
Toys & Games
Game
Love 2 Learn

The Oracle is now the heart of Reading, offering shopping and leisure on three floors and parking for over 2300 cars. The centre includes three major department stores, 28 restaurants, cafés, and bars, and a 10-screen Warner Village Cinema. The River Kennet has been the focus for the centre with bars, cafés, and shops strewn along its banks.

Reading now has branches of the four largest department stores in the UK (M&S, John Lewis, House of Fraser, and Debenhams). Their combined share of the national department store market in 1998 exceeded 70%. Changes in the types of shops in Reading over the last 16 years can be seen in **Figure 5**.

Figure 4 'Shop till you drop' at the Oracle, Reading

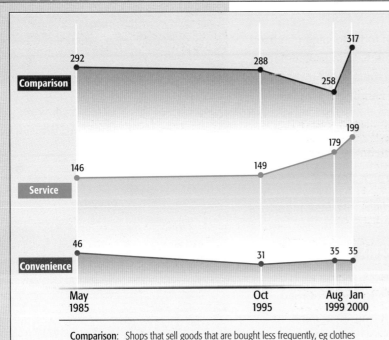

Figure 5 The types of shops increasing or dwindling since 1985 in Reading

Comparison: Shops that sell goods that are bought less frequently, eg clothes
Service: Provides a service to customers, eg. internet café, restaurant
Convenience: Shops that sell 'convenient' goods that are purchased on a regular basis e.g. milk, bread, newspapers

Over to you

1 Compare the two shopping areas (the Oracle and Friar Street) in Reading town centre. Describe what you see.

2 Why might some people prefer the Oracle?

3 In what ways are new shopping centres:
   a better **b** worse than traditional high streets?
   Give reasons for both.

4 How would you persuade new retail companies to move to Reading? Prepare a case to try to persuade one of the following to set up a new store in Reading:
   a Harvey Nichols **b** Harrods **c** Selfridges.

5 How do you think Reading town centre caters for 14–16-year-olds? What new amenities would you like to see? Prepare a case for one of these in no less than 300 words.

EMPLOYMENT

2.6 The M4 corridor

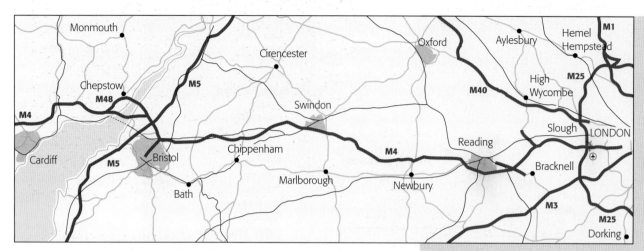

Figure 1 The M4 corridor running from London to South Wales

Figure 2 The M4 motorway at Reading

The M4 runs from West London through Berkshire and Wiltshire and on to Bristol and South Wales. Over the last 20 years it has become an important location for hi-tech industries as well as for many large finance and communications companies. We refer to the term 'M4 corridor' when we talk about the towns along the route of the M4 that have attracted industry over this growth period.

As you have seen in this study companies like Prudential, Compaq, and Microsoft have all located in and around Reading. On pages 42–43 we looked at the advantages of Reading as a town for many of these companies, but many other towns along the route of the M4 have also attracted these 'new' industries such as Swindon, Newbury, Slough, Bristol, and Cardiff. One of the most important factors in the growth of the M4 corridor has been that the 'new' industries are **footloose**.

Footloose means that companies do not need to be located in a fixed place for them to succeed. Raw materials are not an important factor for insurance companies, communications companies or even for many IT companies. Instead these companies find that there are several important factors that need to be considered when choosing a location. These include such things as having an educated workforce, excellent transport links, an attractive location, and cheap land.

Why is the M4 corridor so popular?

The M4 corridor provides many of the advantages that 'new' industries are looking for.

- Land prices are cheaper in towns along the M4 than in London.
- Heathrow Airport is easily accessible from the M4. Many companies find that they need their staff to travel regularly to offices abroad.
- It can be quicker to travel to Heathrow Airport from Reading than from Central London!

Study 2 Reading

Figure 3 Cheap land and nearby universities, not to mention transport, attract footloose companies to the M4 corridor

- There are excellent road and rail links between each of the towns, Heathrow Airport, and Central London. (It takes about 40 minutes to travel into central London by train from Reading. That is quicker than it is by car, train, or tube from Hounslow in West London!)
- Nearby universities at Oxford, Reading, Bath, and Bristol provide an educated workforce and potential for research.
- The sprawling countryside on the edge of these towns provides an attractive environment for offices.
- Many of the towns offer excellent leisure facilities for workers!
- Some companies can benefit from being close to each other, e.g. a microchip supplier and a computer manufacturer. This is called **agglomeration economies**.

Over recent years towns along the M4 corridor have attracted more and more new companies offering higher salaries to the right people. Unfortunately some of these towns are becoming victims of their own success. In Reading it is becoming increasingly difficult for companies to find workers as unemployment has fallen so low. House prices are rapidly increasing and companies are having to pay ever higher salaries to attract workers.

Figure 4 Vodafone have spent £120 million on a new headquarters at Newbury. When completed, 3000 employees will work there

Over to you

1. Describe the location of the M4 corridor.
2. **a** What advantages would you suggest that the M4 should have as a 'corridor' for massive commercial growth, rather than other motorways (e.g. M1, M5, M6)?
   b Produce a sketch map of the M4 corridor between London and Bristol to show these advantages.
3. What problems do you think may arise with the continued success of the M4 corridor?

COASTS

Study 1
Christchurch Bay

For many years, people in the small towns and villages in Christchurch Bay in Hampshire have been fighting the **erosion** of the coast by the sea. They and the local council have tried many schemes, which involve them and the government in great cost. But is the coast worth saving, and what should happen in future?

1.1 How can we manage Christchurch Bay?

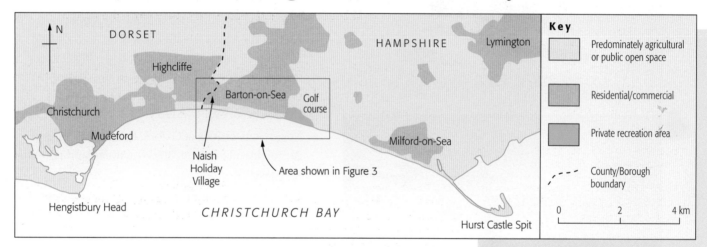

Figure 1 Map of Christchurch Bay

Figure 2 Beach café close to the edge!

Fighting the sea

The small town of Barton-on-Sea has grown close to the cliff-edge. For many years, buildings closest to the cliffs have been under threat. Some have come close to collapse. Sea defences have been built to protect them, but they do not last forever, and need regular maintenance or replacement. The cost is paid for by the local New Forest District Council and the UK government, who pay from local and national taxes.

Figure 3 OS map of Barton-on-Sea and surrounding coast

Scale 1:25 000

54

Study 1 Christchurch Bay

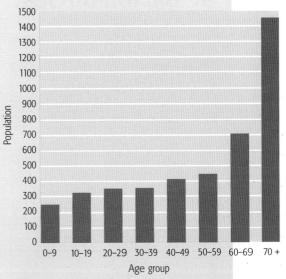

Figure 4 Census data showing the population of Barton-on-Sea

■ Who is involved?

How do local people feel about coastal erosion? Is it something they feel they can put up with, or are they threatened by it? The map of Christchurch Bay (**Figure 1**) shows where people live along this coast. **Figure 4** shows census data about the local population of Barton-on-Sea. Study this carefully and decide why different people have different feelings about the coastline.

You can find out how people feel about issues like this. **Figure 5** is a table of results collected by geography students on a field trip to places around Christchurch Bay. They interviewed some people who lived in Barton-on-Sea and Milford-on-Sea, and some who lived a few miles inland.

Question	Answers from people who live on or very close to the coast (sample of 78 people)	Answers from people who live inland (sample of 33 people)
1 Are you aware of the erosion taking place along the coast?	Yes 83% No 17%	Yes 85% No 15%
2 Are you bothered about the erosion taking place along the coast here?	Yes 78% No 22%	Yes 79% No 21%
3 Does erosion affect you in any way? If 'yes', how does it affect you? If 'yes', what should be done?	Yes 30% No 70% Among the people who said 'yes' were – 'I know people who live there' 'My golf course is being destroyed' 'The beach doesn't look as nice' 'Our local council taxes keep going up'	Yes 7% No 93% Among the people who said 'yes' were – 'I know people who live there'
4 Do you think anything should be done about erosion? If 'yes', what should be done?	Yes 90% No 10% Among the people who said 'yes' were – 'There should be cliff protection' 'A sea wall should be built' 'There should be more groynes'	Yes 79% No 9% Don't Know 12% Among the people who said 'yes' were – 'There should be cliff protection' 'A sea wall should be built' 'There should be more groynes'
5 Who should do the work that has to be done to protect the coast?	Local District or County Council 29% Central Government 68% Charity donations or private individuals 3%	Local District or County Council 33% Central Government 46% Charity donations or private individuals 11% Don't know 10%
6 Would you be prepared to pay extra taxes to pay for any work needed?	Yes 62% No 38%	Yes 39% No 58% Don't know 3%

Figure 5 How people living in places around Christchurch Bay feel about coastal erosion

Over to you

1 Study **Figures 1, 3** and **4** carefully, and make a list of the kinds of people who might live and work here. Give evidence from the map using grid references.

2 Choose four kinds of people from your list. Describe how you think these people might feel about coastal erosion, and why.

3 Study **Figure 5**. In groups, prepare a display to show what local people think of the erosion problem:

   a along the coast   b inland.

Your display should include –

■ graphs to compare people's answers who live along the coast with people who live inland;

■ a comparison to show whether people's answers to questions are similar or different depending on where they live;

■ an explanation to say why people's ideas are the same or different.

1.2 Why is there a problem of erosion at Barton-on-Sea?

Erosion means wearing away of the land surface. When material is eroded from cliffs it is either carried out to sea, where it settles on the seabed, or deposited at the foot of the cliff to form a beach. Not all coastlines erode quickly. Many parts of Britain's coast will change very little over our lifetime. Along other coastlines, however, erosion may be as much as 2 metres per year. Over one person's average lifetime, cliffs can retreat by as much as 150 metres, unless action is taken.

The effects of wave size and direction

Erosion of cliffs is caused by wave energy. Waves are created by wind blowing over the sea surface; greater winds produce larger waves. They often form a long distance away from the coast, and travel a distance known as the **fetch**. The greater the fetch, the greater the wave size. As most surfers know, the largest waves in the UK are found on Atlantic Ocean coasts such as at Newquay in Cornwall.

Coastal headlands may protect coasts from the full attack of waves. In Christchurch Bay, the western bay is protected by Hengistbury Head. However, the eastern bay is open to attack from waves which roll in from the Atlantic. This is the cause of the erosion here. Erosion takes place in two places – at the **cliff foot**, and on the **cliff face**.

Figure 1 Cliff erosion at Barton-on-Sea caused by undercutting and collapse of the cliff

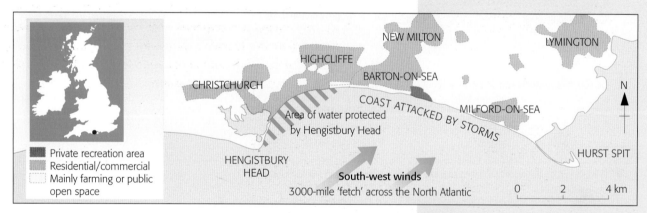

Figure 2 Fetch of waves reaching Christchurch Bay

Cliff-foot erosion

Figure 1 shows where wave attack has eroded the cliff foot. Nearby there are places where the cliff has collapsed. As waves approach the shore, the bed of the sea slows them down. Other waves pile in behind them and rise up, forming a wave which eventually breaks. This attacks the foot of the cliff in three ways:

a) **Hydraulic pressure**. Fractures in rock on sea cliffs contain air spaces. When a wave breaks against a cliff, it forces and compresses air trapped inside cracks. When the wave retreats, air is released with a sudden explosive effect. This fractures the rock, which falls away from the cliff.

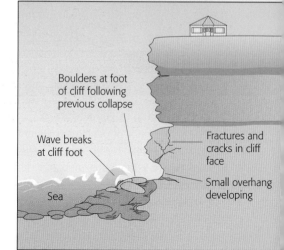

Figure 3 Cliff erosion

Study 1 Christchurch Bay

b) Abrasion or Corrasion. Wave energy throws sand and rock against a cliff as well as water. The action of rock material against the cliff is known as **abrasion**. This cuts into and under the base of the cliff, and causes it to collapse.

c) Corrosion, which is solution caused by the chemical effect of salts or acids in sea water. Weak acids dissolve rocks which contain calcium carbonate, such as limestone or chalk. Salts form on all rocks facing the sea, as salt dries on the surface it weakens rocks already under attack from wave energy.

■ Cliff-face processes

Cliff-face processes include **weathering**, wind action and movement of rocks on the face of the cliff. Together, these are known as **sub-aerial processes**. Weathering disintegrates and causes the decay of rocks at the earth's surface, such as by the action of frost, or by solution and corrosion, described above. **Mass movement** is the movement of rock material downslope, such as cliff collapse, or landslides, which often follow periods of heavy rain. **Figure 4** shows how the cliff has slipped along lines of weakness.

At Barton-on-Sea, mass movement is now the major cause of cliff erosion. It is caused by rain-water soaking into the cliff. Near the surface, the rock type is mainly **permeable** sandy material, through which water soaks easily. This adds weight to the cliff. About 10 metres below, it changes to **impermeable** clay, through which water cannot pass. Where this happens, springs form along the cliff making the cliff face saturated and movement easier. The saturated rocks slump, and this is repeated over time (**Figure 5**).

Figure 4 Slumping along the cliff at Barton-on-Sea

Figure 5 Repeated slumping near Barton-on-Sea

Over to you

1 The following events happen in **cliff-foot** erosion; design a flow diagram to show the sequence in which they occur.

cliff collapses, hydraulic pressure, wave retreats, air is trapped inside cracks, cliff is undercut, rocks thrown with waves against base of cliff, rock shatters, wave reaches base of cliff, an overhang develops, corrasion, explosion of air, wave advances.

2 Using your sequence, show how cliff erosion occurs by drawing a series of diagrams to illustrate the different stages. Use **Figure 3** to help you. Give your diagrams titles.

3 Devise a list of events for **cliff-face** processes. Then re-do questions 1 and 2, this time for cliff-face processes. Label your diagrams and give them titles.

1.3 How can Christchurch Bay's cliffs be protected?

Should local councils protect the coast, or leave it to nature? Along Christchurch Bay, the policy for over 50 years has been to protect towns and villages. Milford-on-Sea and Barton-on-Sea are protected, as well as Hurst Castle further east. How does protection affect people and the environment? Local councils employ engineers to decide which is the best form of protection.

■ Protecting the cliff foot

In some places, the foot of the cliff has to be protected from wave action. This is done by building solid – or 'hard' – features, such as **revetments** (**Figure 1**) and **gabions** (**Figure 2**). Revetments are large boulders placed at the foot of the cliff. The ones at Barton-on-Sea have been bought from Norway – although expensive, they last a long time. Wave energy is thrown against the boulders, and absorbed by them before the cliff can be attacked.

Gabions consist of smaller rocks bound together in a cage of metal. They are best used in protecting beaches. Waves break against the gabions, and the energy is dispersed and absorbed within them. Those in **Figure 2** are also used to protect Hurst Castle further east.

Sometimes, sea walls are built. Compared to other methods, they are expensive, and do not always do the job they were intended to do. When waves attack a sea wall, they rebound; none of their energy is lost. The sea wall at Milford-on-Sea is shown in **Figure 3**.

Figure 1 Revetments at Barton-on-Sea

Figure 2 Gabions used to protect Barton-on-Sea

Figure 3 A sea wall built at Milford-on-Sea

Figure 4 Cross-section through the cliffs at Barton-on-Sea.

■ Protecting the cliff face

The diagram in **Figure 4** shows the rock type – or geology – of the cliffs at Barton-on-Sea. The rocks are divided into different layers, or **strata**. The upper strata of sand are **permeable** – that is, they allow water to pass through. This is a natural process, but causes problems when water reaches **impermeable** clay strata beneath. The clay becomes saturated, then soft and loose, and the weight of the rock above causes the cliff to slump and collapse.

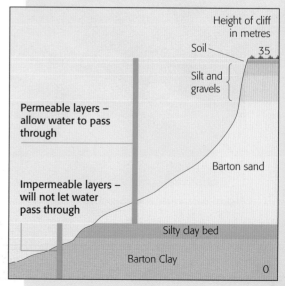

Study 1 Christchurch Bay

To solve this, council engineers have built drains into the cliff and planted shrubs. **Figure 5** shows a drain cover and a trench full of loose gravel and rock, designed to drain water away from the cliff. Drains take water away through channels that have been cut into the cliff, filled with loose rock and gravel.

■ Hard or soft approaches?

Engineers have to make a number of decisions about coasts that suffer erosion. Is a place or a feature worth protecting? If 'yes', then how should it be protected? Those who manage coasts speak of 'hard' and 'soft' solutions to coastal protection.

1 **'Hard' engineering** solutions mean structures that have to be built to alter or to defend the environment. 'Hard' engineering solutions include anything designed to prevent cliff erosion along the shore at the foot of a cliff, such as revetments.

2 **'Soft' engineering** solutions are those that adapt and encourage natural processes to take place. These include:

- off-shore breaks, which are like jetties or reefs built at sea a short way from the coast. Anything can be used – large boulders piled up just above the waterline, or old car tyres, dumped and anchored to the seabed. They are designed to break the force of waves before they reach the shoreline, and reduce erosion;

- beach nourishment, which 'feeds' a beach with more sand in places where erosion is a problem. Sand dredged up from the sea bed can be used to extend the beach, reduce its gradient, and increase friction for waves. Larger, heavier sand particles are used to makes beaches more stable.

Some structures could be considered both 'hard' and 'soft'. For instance, a beach groyne may be a 'hard' built structure, but it encourages a natural beach to develop.

Figure 5 Drainage on the cliffs at Barton-on-Sea

Over to you

1 Copy **Figure 4** and add extra details and labels to show how cliff collapse can occur. Number your labels 1, 2, 3 etc. in order to show the stages in which it occurs. Give your diagram a title.

2 Now make a second copy of **Figure 4**. On it, show where drains are dug, and use labels to show how they reduce cliff collapse. Give your diagram a title.

3 Decide which of the following are hard and which are soft solutions to the problem of coastal defence; some may be both. Give reasons for your choice in each case.

Sea wall, cliff drains, a jetty built out to sea, dredging sand from the sea-bed on to a beach; revetments; gabions; leaving the coast to natural processes.

1.4 Solving one problem, creating another?

Beach protection

Christchurch Bay has suffered gradual erosion of its beaches. Over the years, beach sand and pebbles may be moved several kilometers along the coast. Along the bay, waves break on-shore from the west and move material, shown in **Figure 1**. As they break, they rush up on to the beach at an angle, moving beach material from A to B. Water then drains back down the beach slope towards the sea, taking beach material with it to point C. As the next wave breaks, the same beach material is moved from C to D, and back down the slope to E, and so on. This process is known as **longshore drift**.

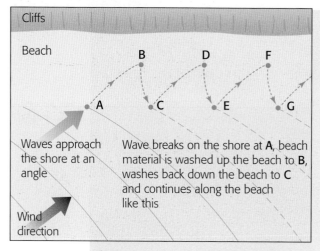

Figure 1 The process of longshore drift

Along the coast, beach material has been carried east to the Solent estuary to form a sandy **spit**. A spit is a long peninsular of sand created by longshore drift. **Figure 2** shows how it has been formed. At Hurst Castle, it meets a current of water bringing river water out to sea. Water currents along the Solent at high and low tide shape the spit, curving it inland so that it is shaped like a claw. Behind the spit, a **salt marsh** of mud and sand has formed.

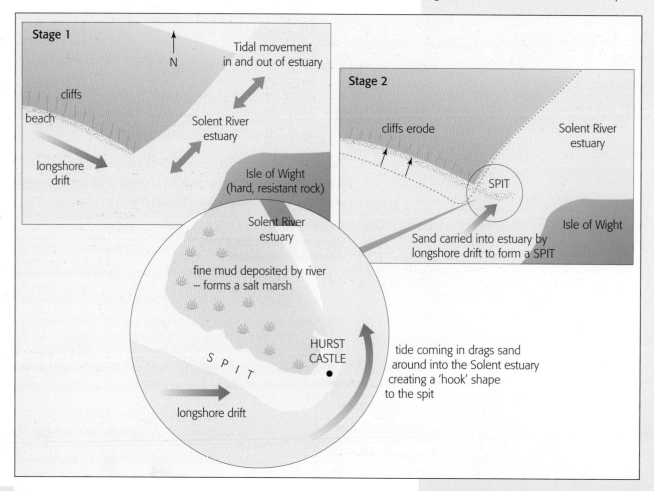

Figure 2 The formation of Hurst Castle spit

Study 1 Christchurch Bay

■ Using groynes to protect the coast

Engineers now know how important longshore drift is in protecting cliffs. Beach material builds up and acts as a break to wave attack, absorbing wave energy before it reaches the cliff. The very thing that protected the cliff at Barton-on-Sea in the past was its beach. Much work has been done to allow a build-up of sand, using **groynes** (**Figure 3**). A groyne is built at right-angles to the beach, and can be built of wood or stone. **Figure 3** shows how sand is trapped by the groyne from moving further east along the shore. This process of protection is known as **beach nourishment**.

But the construction of groynes has led to problems elsewhere. **Figure 4** shows the cliffs near to Naish Holiday Village to the west of Barton-on-Sea. In the photo, waves move sand from west in the distance to east in the foreground. In the distance, groynes have been built to protect the beach from erosion. Since these were built, no new sand is brought by longshore drift to replace that taken away. The result is that the coast has eroded rapidly in ways shown in the photo.

Figure 3 A stone groyne at Barton-on-Sea

Figure 4 Erosion near Naish Holiday Village in Christchurch Bay since groynes were constructed further west

Over to you

1. Refer back to **Figure 3** on page 54.
   a. identify where groynes have been built along the coast at Barton-on-Sea, and give 4-figure grid references for these.
   b. Suggest why these have been built at Barton-on-Sea, but not at Naish Farm further west.
   c. How might local residents feel about this, and why?
2. Imagine that you are on work experience at New Forest District Council. You are helping engineers to decide how to spend a small budget for coastal protection at Christchurch Bay. This year, they have only enough to choose **two** out of revetments, gabions, groynes and cliff drains.
   a. Copy the table below and complete it, to show as many advantages and disadvantages of each as you can think of.
   b. Recommend which ones you think are best for protecting Christchurch Bay.

	Advantages	Disadvantages
Revetments		
Gabions		
Groynes		
Cliff drains		

1.5 Should Hurst Castle spit be protected?

The cost of engineering is high. Most costs are met by the Ministry for Agriculture, Forestry and Fisheries (MAFF) to whom local councils have to bid for money for coastal defence. It now costs so much to protect the coast that MAFF has had to re-think its ideas about coastal defence. 'Soft' solutions to coastal defence are cheaper than 'hard'. This means that some cliffs might have to be left to erode naturally, or be protected by cheaper options such as beach nourishment.

Decisions are easy to make when villages or towns need protection. But what happens when an environmentally valuable area needs protection? Hurst Castle spit at the eastern end of Christchurch Bay provides an example.

Figure 1 Hurst spit showing the location of Hurst Castle

Figure 2 The salt marsh behind Hurst Spit, now an SSSI

Hurst Castle lies at the end of Hurst Spit, about 2.5 km from the end of Milford Beach. Behind the spit lies a large salt marsh, which is a **Site of Special Scientific Interest** (SSSI). The salt marsh is a habitat for a variety of wildlife, especially birds, invertebrates, and plant life. This is known as its **ecology** – that is, the plants and animals that exist together. Some of its features are shown in **Figure 3**. At the end of the spit, the castle is an Ancient Monument and was built by Henry VIII. If a storm were to breach the spit, the SSSI would disappear rapidly. Storms would also reach much further up the Solent towards Southampton.

Study 1 Christchurch Bay

Figure 3 The value of Hurst Spit

Leisure
The spit protects the Solent from strong winds, making sailing safer. There are two sailing clubs locally. The spit is popular with walkers, fishing enthusiasts and naturalists.

Protecting the Solent
Hurst Spit takes the full force of storms and keeps the Solent estuary calm. This makes it safer for shipping, and protects towns on the Solent.

Plant life
Although the spit is mainly shingle and sand, where little will grow, the salt marsh is a site of special scientific interest (SSSI). The marsh plants, such as reeds and sea lavender, are unusual as they have to live above water for half the day and below it for the other half. So many estuaries have been taken over by human activity (e.g. the Thames, Tees) that few other estuary plant communities like this are left.

History
The castle was built by Henry VIII to defend England against Spain and France. It has since been extended and played a key part in defence in the Napoleonic Wars. A lighthouse was added in 1786. Now it is run by English Heritage and is open to the public in summer. There are also old salt workings from the nineteenth century.

Insects

Birds
Birds here, including the colourful Oyster-catcher, Ringed and Grey Plovers and Redshanks, all feed intertidally and nest on the salt marsh. Dunlin outnumber all other waders with about 5000 individuals. They are hard to spot, but fascinating to watch looking for food on the mud following the ebb tide. More conspicuous are the Black-headed Gulls with about 1400 individuals in the summer nesting on the salt marshes between Hurst Castle and Lymington, with a similar number in the Beaulieu estuary. Together, these form one of the largest colonies in the UK and almost 1% of the world population. They depend on this as a winter resort.

Over to you

1 Draw a sketch of **Figure 1** on page 57, labelling the following features:
   Hurst Spit, Hurst Castle, the salt marsh, the sea wall protecting the castle, Milford-on-Sea, groynes.

2 **a** Describe the area around the spit, using the information and Figures on pages 60–63.
   b Describe the shape of the spit, the beach, the marshes.
   c Show by arrows the direction of longshore drift along the spit, and how it has been formed.

3 **a** Why should there be a danger of breaching the spit at point X on **Figure 1**?
   b What could be the effects on the spit, the salt marsh and the castle if the spit were breached? Label these effects on your map.

4 Is Hurst Castle worth protecting? Make a copy of the following table and decide on the advantages and disadvantages of protecting it.

Advantages of protecting Hurst Spit	Disadvantages of protecting Hurst Spit

5 Consider the following organizations that have some say in coastal management in this area.
   - UK government – responsible for paying 70% of costs of coastal protection
   - New Forest District Council – responsible for designing coastal protection
   - English Heritage – responsible for managing Hurst Castle
   - Environment Agency – responsible for the quality of water and the environment
   - Royal Society for the Protection of Birds.

   In pairs, decide whether these organizations would support the protection of Hurst Spit or not, and why. Write up your decisions.

1.6 Is the cost of protection worth it?

By now, you will know how coasts are managed; how decisions have to be made about which methods to use; whether those methods are worth the cost.

Finally, we will consider whether money should be used in protecting the coast, or whether it should be left to the forces of nature.

How effective has coastal protection work been so far? **Figures 1** and **2** show an aerial view of Barton-on-Sea, one from 1967 and the second from 1997. Use these to identify whether or not erosion has slowed down in 30 years. If it has, it is worth doing; if not, should more money be spent?

Sometimes local councils decide to let natural processes do their job. Rather than leave the coast to erode rapidly, they decide on a policy of **managed retreat**. This means that they might either:

- allow spending on beach nourishment using groynes, but not spend huge sums on expensive revetments, or
- decide to spend no more.

Figure 1 Barton-on-Sea in 1967

Figure 2 Barton-on-Sea in 1997

Study 1 Christchurch Bay

Figure 3 The coast near Naish Holiday Village

Figure 3 shows the coast near Naish Holiday Village just west of Barton-on-Sea. In recent years, erosion there has been very rapid – about 2 metres per year. Unprotected by a beach, waves move onshore, and erode the base of the cliff. Without the drainage works found at Barton-on-Sea, slumping is frequent. Should it be protected? While the cliffs are a danger, and the beach is so poor, tourists may stay away. If a programme of protection were carried out, more tourists might visit the area and help the economy.

What do you think should be done?

- **The cliffs at Naish Holiday Village are of international geological importance … they contain an outcrop of fossil-bearing strata. Continued exposure through erosion is part of their importance.**
- **The beach is popular with summer visitors, but unstable cliffs prevent safe access from the cliff top.**
- **Despite the supply of fresh material from the cliffs, the beach is unstable. The cliffs also result in a poor environment of low landscape quality.**

Figure 4 Extract from the New Forest District Coastal Management Plan 1997

Over to you

Study **Figure 1** and **Figure 2** carefully. Identify housing in Barton-on-Sea, the road running parallel to the coast, the cliff line, slumping along the cliffs, the beach, and beach groynes.

1. Find features A,B,C,D and E on **Figure 2**.
   a Identify the features in each case.
   b Say what changes have taken place since 1967.
   c Suggest reasons for each change.
2. If cliff erosion were to continue, estimate how long would it take:
   - for the car park at **E** to disappear?
   - for the cliff line to reach the first house nearest the cliffs?
3. Draw a table to show 'Features that have disappeared from Barton 1967–1997' and 'New features in Barton-on-Sea since 1967'. Which list is the longer? Why?
4. Why do you think that the council has continued to allow building here?
5. Discuss in pairs whether you think that coastal protection here has been successful or not.
6. Decide, with reasons, whether you think that the local district council should now:
   a allow erosion to continue without any further protection
   b save Naish Holiday Village
   c carry out more engineering work for greater protection
   d set up a programme of beach nourishment to improve the beach.

COASTS

Study 2
The Daintree World Heritage Coast

This section is about part of Australia known as the Daintree, an area of northern Queensland. It shows how some coastlines are considered to be worth protecting, and how they need to be managed to cope with economic development.

2.1 Another day in paradise?

Figure 1 shows the location of the Daintree. The Daintree is the area of coast shaded in on Figure 1, and is part of a huge stretch of rainforest that runs parallel to the Queensland coast, from Cape York in the north-east of Australia for about 1000 miles southwards. Until recently, it has been difficult to reach the Daintree. Many of its beaches are undeveloped, and fringe the coast which faces directly on to the Great Barrier Reef. From almost anywhere for about 1500 miles, the Barrier Reef is only a few hours' sailing. Here, diving, snorkelling and a wide variety of water and beach environments attract people from all over the world. At about 16 degrees south of the Equator, temperatures never move much from about 28°C all year. Paradise!

Because of its environment, the area now attracts many more tourists than it did. Tourism is now competing with many other economic activities. The coastline is changing as more development takes place. This section looks at some of the developments taking place and asks whether the coast should be protected further.

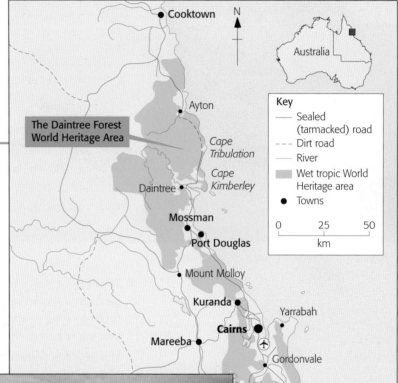

Figure 1 The Daintree coast of northern Queensland

Figure 2 The northern Queensland coast

Figure 3 The climate of northern Queensland

Annual average temperature: 24.8°C
Rainfall total: 2025 mm

66

Study 2 The Daintree World Heritage Coast

■ The growth of tourism

Tourism is the fastest growing industry in the world, and especially in Australia. At present, the northern Queensland coast is long enough to take many of the people who visit. But pressure on the environment is growing. In 1999, Australia attracted about 4 million visitors from overseas. The majority of these tend to visit the large cities of Sydney (59%) and Melbourne (24%). In these two urban areas live nearly half of the country's 19 million people.

The number of visitors to Australia is increasing rapidly. Three reasons explain this:

a) Overseas air travel is now much cheaper than it was 10–15 years ago. In 1985, it cost about £850 for a low-season fare to Australia from the UK. In 2000, fares are being advertised for the same season for £600.

b) In 1985, the fastest service by air from London to Sydney took 25 hours. In 2000, it took 21 hours. In 1985, Qantas, the Australian airline, and British Airways each had one flight daily to Sydney. Each airline now has two flights a day to Sydney alone, plus others to Melbourne, Brisbane, Adelaide, and Perth. Over 20 airlines now fly from the UK to Australia.

c) The Australian dollar has weakened as a currency, and is now relatively inexpensive to overseas tourists. Eating out is about 30% cheaper than in the UK.

Now, the third largest destination of overseas visitors to Australia is northern Queensland. Until recently, this area has been isolated from the rest of the country by distance and cost. The distance from Sydney to Cairns, the largest town in Northern Queensland, is 1600 miles (2500 km) – nearly as far as from London to Athens! To drive, it would take 3 or 4 days. As a result, most people fly there, and Australian internal air fares were not cheap until recently. This has helped to protect the coast from mass tourism.

	1985	2000
Cost of discounted return fare London Heathrow – Sydney by Qantas during August	£850	£615
Time taken by fastest flights	25 hours 15 min	21 hours 30 min
Number of stops en route	2	1

Figure 4 Journey times and costs of travelling to Australia from the UK, 1985–2000

Country of origin	Number of visitors to Australia in 1997
Japan	766 000
New Zealand	621 000
UK	387 800
USA	309 800
Korea	220 500
Singapore	201 300

Figure 5 Numbers of overseas tourists to Australia from different countries, 1997

Over to you

1 Describe the location of the Daintree area, using **Figure 1**. Mention the country it is in, the state, the part of the state and its distance and direction from the nearest city and airport. Use your atlas to describe its position north or south of the Equator.

2 Using **Figure 2**, think of 20 words to describe the things you see and your feelings about what you see. Use these words in a paragraph to describe the photo.

3 Why should this area have become popular with tourists from overseas and within Australia?

4 Using **Figure 3**, draw a climate graph for northern Queensland. Give your graph a title. Then answer the following.
   ■ When are the warmest and coolest months? Label your graph with summer and winter seasons.
   ■ When are the wettest months? Is this summer or winter?
   ■ How do you think Australians here describe the seasons of winter and summer? How do they compare with the UK's winter and summer?
   ■ How might this climate be a factor in explaining the growth in tourism?

5 Explain how the data in **Figure 4** have made Australia much more accessible to overseas visitors during your lifetime.

6 On a world map, draw flow lines to show where tourists to Australia come from (**Figure 5**).

7 Find out about the Great Barrier Reef, using an encyclopaedia such as 'Encarta' or 'Encyclopaedia Britannica' or by using the internet. What is it? Where is it? How big is it? What does it consist of? How has it developed? What are the issues it faces?

2.2 What are the pressures on the Daintree?

The Daintree coast has become valued as one of the world's most beautiful and unspoilt areas. It is now a **World Heritage Site**, created by the Australian government in 1987 so that it would be protected from further development. To get it to that stage took a great deal of fighting, however, and the battle for the Daintree is not yet won.

Tourism is now putting pressure on the northern Queensland coast, and economic development is rapid. There is evidence to suggest that planning is needed now before development gets out of hand. A number of special environments are at risk. Why do tourists come here, and what pressures do they put on the coast?

Figure 1 Rainforest along the Daintree coast

Figure 2 The strangler fig, a rainforest plant in northern Queensland

What's special about the Daintree?

a) Scenery and rainforest

Figure 1 shows the rainforest in the Daintree area of northern Queensland. The landscape is a fairly unique combination of coast and rainforest, with mountains set behind the coast. The mountains attract heavy rain for most of the year, which has given rise to lush rainforest. Until recently, the Daintree forest has remained relatively untouched. Many tourists visit the area to see the fauna – such as butterflies and bird life, with parrots and cockatoos seen freely in the wild – and flora, with typical rainforest plants.

Tropical forests contain a huge range of plant and animal species. Approximately 155 000 of the 255 000 known plant species have been found here. This is probably only a small proportion of plants thought to exist in the world; estimates vary between 5 and 30 million. Less than 1.5 million of these have been properly described and identified scientifically.

There is a huge range and number of plant and animal species in rainforests. This is called **biodiversity**. A high proportion of the world's bird and primate life is found in these areas. One-fifth of all bird species are found in Amazon rainforests alone, and 90% of all primates are found only in tropical rainforest areas. Not only is this of value in the ecology of the world, but there are uses for people as well. Natural genes from rainforest species are used to improve resistance of crops to pests or disease. This biodiversity is under threat. It is thought that already 10% of the world's species had become extinct by 2000.

Study 2 The Daintree World Heritage Coast

The Daintree forest in northern Queensland is considered very special by ecologists. Over many millions of years, it has been virtually undisturbed. As a result it contains species of plants that are found only in northern Queensland, and nowhere else. It contains some spectacular examples of rainforest plants, such as the strangler fig shown in **Figure 2**. This fig is a parasitic plant – that is, it survives living on others – and engulfs a host tree around which it grows, living on its sap, and finally killing it. Other plant species are found that are normally accepted as house plants in the UK, such as Ficus, Swiss Cheese plants or Yucca. Many species of palm, and hardwood trees are also found. Some animal species are also rare, like fruit bats, as well as bird life.

b) The Barrier Reef

The Barrier Reef is located off the east coast of Australia, and runs for 1000 miles parallel to the coast of Queensland. It consists of large numbers of coral reefs, which are among the world's largest living water ecosystems, growing in shallow, clear water and supporting huge varieties of fish and marine life (**Figure 3**). Coral reefs consist of billions of coral polyps, which live together in reefs, or colonies. Each coral polyp is an animal, and belongs to the same group as jellyfish and sea anemones. Each is about 2 to 3 cm in length and feeds on minute organisms in the sea.

Coral reefs are a vital part of the global ecology. They support 25% of marine species, and protect coastlines that would otherwise suffer from erosion during storms, and act as natural recycling agents of carbon dioxide from seawater and the atmosphere. Along the Daintree coast, the Barrier Reef acts as breakwater to the rest of the coast, providing calm water inside the reef. The reef is delicate; corals break easily and are soon destroyed, so much effort is made to protect the coast. A few operators are granted the right to run boat trips such as the one shown in **Figure 4**, which operates out of Port Douglas. Tourists can travel for a day's diving, snorkelling and underwater reef viewing. Their activities are monitored to make sure that damage is minimal, paid for by local tourist taxes.

Figure 3 Underwater view of the Barrier Reef

Figure 4 One of the 'Quicksilver' boats, which takes tourists from Port Douglas on a 90-minute trip to the Great Barrier Reef for a day's diving, snorkelling and underwater reef viewing

Over to you

1. Write about 150 words describing the value of tropical rainforests in northern Queensland, as seen by an environmentalist. What is it about them that you most value?
2. Now write a further 150 words about these rainforests as seen by an economist, who would view them for their economic potential. What is it about them that you most value?
3. Now complete your writing, with a final 150 words about tropical rainforests, as seen by a tourist visiting them. What is it about them that you most value?
4. Now do the same exercise for the Great Barrier Reef.
5. Would you say that tourists are more likely to side with environmentalists or economists in their views about the rainforest and the Barrier Reef?

COASTS

2.3 Are we destroying what we most want to see?

Who are the tourists?

Until recently, the tourist market in northern Queensland was quite specialized, and appealed only to a few. Although Cairns has several large resort-style hotels, most tourist investment further north was limited. Most tourists have been backpackers, or travellers who have taken a year (or more) out from universities, college, or employment. They tend to be attracted to the area because it remains largely unaffected by mass development. Backpackers are valuable to the Australian economy. They spend four times longer in the country than other tourists and spend an average of $4000 during their stay. They tend to spend money in the shops, pubs, and restaurants as a result, and this helps to create plenty of local employment. They also contribute to local employment, forming a valuable pool of informal labour, able to work part-time hours for short periods in ways that employers like.

Now investment is changing.

- Japanese investment here has resulted in several golf courses, targeted largely at the south-east Asian market.
- The Sheraton and Radisson international hotel chains have developed large resorts in Port Douglas, further south.
- A number of Hollywood films have been made here, with stars such as Tom Cruise and Sean Penn helping to publicize the area to a more 'up-market' clientele.

As a result, places such as Port Douglas, below, and the Daintree are changing.

Growth and change in Port Douglas

One place that is increasingly affected by development close to the Daintree is Port Douglas, a small town to the south of the Daintree, shown in **Figures 1** and **2**. Although the population of Port Douglas is only about 2000, it is rising rapidly, and expanding as more accommodation is developed for tourists. In recent years, visitor numbers have grown rapidly, and change is affecting the character of the village. The amount of accommodation developed recently to house tourists is shown in **Figure 4**.

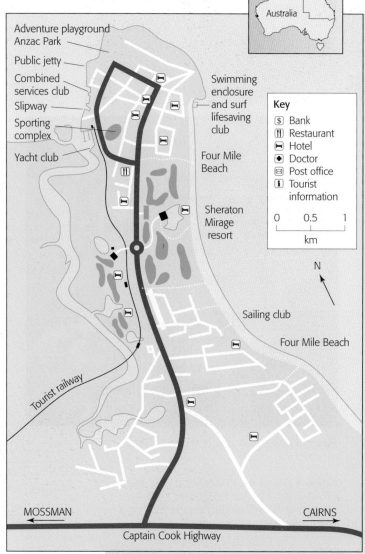

Figure 1 The location of Port Douglas on a coastal spit

Figure 2 Port Douglas and its 'four-mile beach'

Study 2 The Daintree World Heritage Coast

Figure 3 The traditional village centre in Port Douglas

	Dec 95	Dec 99	Completed by 2000
Apartment	568	994	281
Hotel	782	1121	626
Motel	83	83	0

Figure 4 Different types of accommodation in Port Douglas 1995–2000. The figures show the number of bedrooms available in holiday apartments, hotels and motels.

The number of apartments and hotels has led to other changes, caused by the increase in the number of visitors.

A large supermarket was built in the village centre in 1999, and has changed the way that people shop. Local small shops complain that people no longer shop with them.

In 2000, the local council gave permission for a new McDonalds to be built on the site shown in **Figure 5**.

Some people are concerned that changes like this are altering the character of Port Douglas. A property 'boom' has led to rises in housing prices, which are affecting local people. Some people have done well out of this, selling their land to property developers. Other people, such as young families, now find it difficult to buy a house or apartment.

Will the development spread north?

With so much change in Port Douglas, people now wonder if it will stop there or spread to Daintree. Although there has been little building in the Daintree area so far, some people worry that it is only a matter of time before hotels, restaurants and shops spread further into the rainforest.

Figure 5 The site for a new McDonalds restaurant in August 2000

Over to you

1 Write out a title 'Change in Port Douglas'.
- Briefly describe its location, and in a sentence how it is changing.
- Identify amenities and services available in the town.

2 What is the difference between resorts, apartments, hotels and motels? Write a sentence definition of each term.

3 Draw a graph to illustrate the change in number of apartments, hotels and motels in Port Douglas in 1995, 1999, and 2000. Give your graph a title.
  a Is the graph showing faster, slower or no growth? Explain your answer.
  b Explain possible reasons for the changes.

4 What might the changes in numbers of rooms available mean for Port Douglas:
  a now, and   b in the future?

5 You have been asked to produce a 300-word extract for the 'Rough Guide' to Port Douglas and the Daintree coast for people who are planning a gap year after the sixth form. Describe the area, and say whether you think it would appeal to 18–19 year-olds, and why.

6 Use Studies 2.2 and 2.3 (pages 68–71) to summarize:
  a how Port Douglas and the Daintree are changing
  b how changes are affecting the area
  c whether you think changes are for the better or worse.

COASTS

2.4 It's not just for the tourists

In recent years, economic development along the Daintree coast has led to conflicting views about what the coast should be like in future. Some wish to protect it, while others believe that it should be opened up for economic expansion. Environmental groups are concerned to protect the 'special' nature of the Daintree. Several protests have taken place against development.

- A protest took place in the late 1980s about the extension of a road northwards through the Daintree; protestors felt that it would bring unwelcome change to the coast and rainforest environment;
- A protest in Cardwell, further south near Townsville, threatened the development of Port Hinchinbrook in 1999.

The fear of environmental groups is that development will take over the coast. They believe that the coast is fragile and needs protection. On the other side, developers see opportunities for land which is undeveloped and holds plenty of scope for farming, tourism and industry. They feel that the remoteness of the area needs an economic 'boost' to bring more employment for local people. Besides tourism, there are several types of employment here focused around farming and timber.

Farming

Farming has been the main economic activity along this coastal region for over 100 years. Although the sugar industry has been hit by falling world demand for sugar, it still occupies large estates consisting of huge fields of cane. In addition, the climate is ideal for growing a range of tropical fruit, and mangoes, pineapples and banana plantations (**Figure 1**) are common. These crops support other industries, such as the sugar mill at Mossman (**Figure 2**).

Timber

The timber industry would dearly love to cut the timber in the forests. They argue that less than one-fifth of rainforest in Queensland is used for timber and that there is still plenty to cut. In the past, rainforest has been cut to make way for large sugar-cane farms. The timber companies say that nobody complained then. They support those who want to build and extend the road. They say that environmentalists who want to protect the forest are depriving people in Queensland of jobs.

Home-based industries

As well as large employers, such as the sugar factory, the Daintree has attracted many small, home industries, run by people who were originally attracted to the Wilderness quality of the Daintree. Many 'alternative' lifestyles have developed here, with industries such as organic farming and pottery-making.

Figure 1 Banana plantation near Mossman

Figure 2 Sugar factory at Mossman

Figure 3 Cattle farming in the Daintree

Study 2 The Daintree World Heritage Coast

■ Managing conflict in the Daintree area

Many issues raise all kinds of feelings. Some people object to development in the Daintree area because they believe that the environment is special. Others feel that it should not go ahead because they already live there and do not wish to see it spoiled by large-scale development. They think the same thing but not for the same reason.

It is possible to show how different people feel about an issue by plotting a **conflict matrix** like that shown in **Figure 4**. A conflict matrix identifies anyone who may have a feeling about a development – for or against, and lists them down the left-hand column in the rows of the table.

The same people are shown in the other columns of the table.

Each person in each row is then judged in terms of how much they agree or disagree with each person in each column. For instance, in the example in **Figure 4**, sugar cane farmers are likely to agree with workers at the sugar mill in Mossman about whether the area should be developed further, because their jobs and wealth depend upon sugar.

The completed table will show who is likely to agree or disagree, and by how much, over an issue.

	Sugar cane farmers	Workers at the Sugar mill in Mossman	Backpackers	Environmental groups	Local residents in the Daintree area
Sugar cane farmers					
Workers at the sugar mill in Mossman	+ + +				
Backpackers					
Environmental groups					
Local residents in the Daintree area					

Figure 4 A conflict matrix

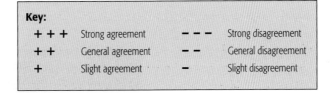

Key:
+ + + Strong agreement   – – – Strong disagreement
+ +   General agreement   – –   General disagreement
+   Slight agreement   –   Slight disagreement

Over to you

1 Make a copy of the conflict matrix in your book. Then:
- Think of any other people who would have an opinion about whether or not there should be further development along the road into the Daintree.
- Write their names into the left-hand column of the matrix, and again at the top of a column on the right. You can add as many as you like.
- Decide who agrees with who, and how much, by using + or – symbols as shown in the key.

2 In a few sentences, explain what the table is about and how you drew it.

3 Describe what the table shows about who agrees, and who disagrees with further economic development.

4 Select any two people who strongly agree, and two who strongly disagree. Suggest why these people agree or disagree with each other.

5 What do you think are the main arguments for:
   a opening up the rainforest
   b protecting it from the development?

COASTS

2.5 Which way now? – 1

You have seen how different people use this coast for different purposes – tourists, farmers, and loggers. At present the world's rainforests are disappearing rapidly. By 2020, it is feared that the largest forest in the world, the Amazon, will have all but disappeared. What should happen to the Daintree coast – should it be developed, or protected for the future?

How is the Daintree changing?

Much of the Daintree coast is still rainforest wilderness and difficult to get to. Until 1985, there was no road north of the Daintree River. Even after the completion of the road, Cape Tribulation (see **Figure 1**) was accessible only by 4 wheel drive. But bringing people into the area changes its character. How should the future Daintree area look? Two possible alternatives are shown here and in the next study:

- to develop the area economically
- to develop sustainably – that is, to try to keep the character of the area and place strict limits on how much development takes place in order to preserve it for the future.

Option 1 Go for economic development!

The economic benefits from tourism are great. In 1983 about 17 000 tourists visited the Daintree. By 1999 more than 300 000 people came there, bringing an industry worth AU$100 million ($40 million). Although the area is a long way from mass tourism, Figure 1 shows that there are already many amenities along the road, such as new shops and restaurants. Many would like to open up the Daintree for its potential. **Figure 2** shows two photos; one from 1992, the second from 2000. Between these years, the road had been sealed, or tarmaced. Now lorries and buses are able to bring goods and people into the area. Already, small areas of forest have been divided up into plots for sale (**Figure 3**).

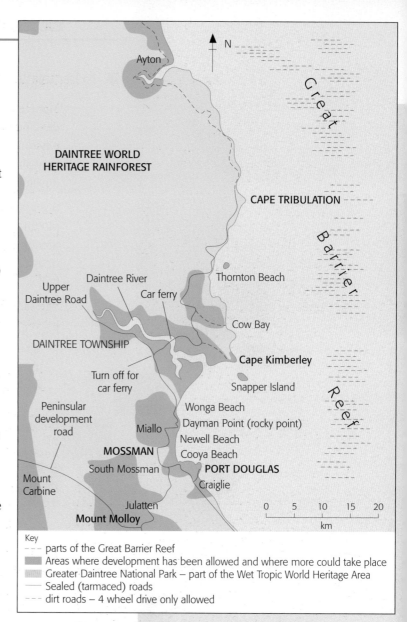

Figure 1 The location of Cape Tribulation. Notice how the roads leading to it require 4wd vehicles. The map also shows amenities and services along the road to Cape Tribulation.

Figure 2 The road through the Daintree in a) 1992 and b) 2000

Study 2 The Daintree World Heritage Coast

Figure 3 'For sale' signs like this are appearing in forest plots as people and developers begin to move in

Figure 4 The Daintree ferry

There are three main limits to development at the moment.

- The ferry crossing the Daintree River (**Figure 4**) limits traffic levels. Although frequent, the ferry can only take about 20 vehicles on each crossing.
- There is no mains electricity north of this point. Anyone who wants to move here has to provide their own generator or solar-power system.
- The level of services can only support a small number of people – there is no mains water, no mains sewerage disposal system, and few shops and services for local people.

Some people would like to achieve the following.

- Replace the ferry across the Daintree River with a bridge in order to increase the visitor limit in the area.
- Extend mains electricity cables into the area.
- Install services that can support larger numbers of people, such as mains water, and sewerage systems.

Sometimes it helps to see what can happen when development is allowed to take place elsewhere. Then we can ask the question – 'do we want this to happen here?' People in the Daintree area look further south to Cardwell, a small town south of Cairns. There, the development of Port Hinchinbrook has caused a major split in townspeople. You can see this in programme 2 of BBC's 'Australia 2000' series.

Should this be the kind of development to take place? People in Cardwell use the same arguments and have the same hopes and fears as those in the Daintree further north. **Figure 5** gives you some idea of what they can see happening.

Figure 5 Temptations for the future – a new resort at Port Hinchinbrook

Over to you

1. Make a copy of the table below. Think about the three schemes shown, and complete the table with as many advantages and disadvantages as you can think of.

Scheme	Advantages	Disadvantages
1 Replace the ferry across the Daintree River with a bridge		
2 Extend mains electricity cables into the area		
3 Install services such as mains water, and sewerage systems		

2. Using three highlighters or coloured pens:
   - mark in one colour any advantages and disadvantages that affect people (social)
   - in another colour those that affect the economy (e.g. work)
   - in a final colour those that are environmental.

3. **a** In pairs, discuss the following statement:

   'The advantages of the three schemes for the Daintree would be economic, but the disadvantages would be environmental'.

   b Using examples, write about 200 words to say whether you agree or disagree with this statement.

4. Do you think that the development at Port Hinchinbrook (**Figure 5**) is an argument that supports further development of the Daintree or rejects it? Explain your answer.

COASTS

2.6 Which way now? – 2

■ Option 2 Go for conservation!

Many people have protested about further development in the area. Those who want to save the area say that they have strong arguments on their side.

1 Cleared forest will lead to greater run-off each time it rains, which would erode soil from the land. The soil would wash into rivers, drain out to sea and disturb the clear waters of the Barrier Reef. **Figure 1** is a photo taken only weeks after the dirt road was opened in 1985. Already you can see that the soil is eroding from the cutting.

2 The threat to biodiversity. It is thought that, worldwide, 10% of the world's animal and plant species in rainforests had already become extinct by 2000.

They support the area's World Heritage status, and want to keep it for the benefit of people not only now but in future. This kind of policy is known as a '**sustainable**' one – where what happens now helps to maintain and protect the area for future use. They want to prevent anything that works against a sustainable future for the Daintree, such as:

- environmental destruction – e.g. pollution; or damage to the environment or which destroys rare species.
- using unnecessarily those resources that cannot be renewed.

Their concern is the area between the Daintree River and Cape Tribulation. Now that the road is tarmacked as far as Cape Tribulation, they want no further development.

The Daintree area is fragile. One species – a member of the red cedar family of trees – became extinct in 2000 when a landowner cleared a forest plot. If land clearing is not reduced in the Daintree area, 85 rare plant species now on private land will also become extinct. Now, a group known as the Rainforest Co-operative Research Centre has put proposals to the local council in Port Douglas, based upon a study that it carried out, shown in **Figure 2**. It recommends much lower levels of development, following a study of the local environment and population.

Figure 1 Road cutting along the road to Cape Tribulation – notice the soil eroding from the sides

Figure 2 Report from the 'Cairns Post' 16 August 2000

REPORT URGES DAINTREE LIMIT

By Nick Dalton

THE number of people living in the area north of the Daintree River should be limited to 1200, a report released late last night says.

The Rainforest Co-operative Research Centre report says about 500 people currently live in the area.

But it said the local Council planned to allow 2400 people to live there.

The report also recommends
● Limited future residential development.
● Nature-based tourism and associated businesses.
● Organic farming on cleared areas.
● Provision of electricity through a combination of limited overhead power lines and diesel generators and solar power.

The report concentrated on future possibilities for the 1200 residential allotments of privately-owned land.

There have been disagreements about electricity supply, tourism, development and tourist visitor numbers.

The project aimed to look at ways of creating an ecologically-sustainable future for the Daintree community at the same time as combining conservation with needs of residents and interests of tourism.

Study 2 The Daintree World Heritage Coast

The study recommends that the local council should:

- Limit the population to about 1200;

- Force people to use **renewable energy** supplies, such as solar power, which they would have to install at personal expense if they wanted to settle in this remote area;

- Buy back land purchased by developers who had hoped to build and develop on it;

- Encourage people who have already bought land not to use it, but keep it as it is. This would discourage people already there from clearing more land. It could even include fines if an owner clears any more land, especially in areas of endangered species. Fines could be as heavy as taking land away from owners who refuse to comply;

- Create a world-class scenic hiking trail that would include spectacular forests, secluded beaches and some of the most rare and endangered plants in the world.

Their views differ from those of the local council, who want more development in the Daintree. The local council includes farmers and land developers who want a larger scale of development. They want about 2400 people at this stage, and perhaps more later.

Figure 3 The Daintree – a landscape and ecosystem worth protecting? Imagine a scenic hiking trail through this area

Over to you

1. Make a copy of the table below. Think about the five proposals, and complete the table with as many advantages and disadvantages as you can think of.

	Advantages	Disadvantages
Limit the population to about 1200		
Use renewable energy supplies, such as solar power, at personal expense to those who settle in this area		
Buy back land already purchased by developers		
Encourage people who have already bought land not to use it, but keep it as it is		
Create a world-class scenic hiking trail, which would include spectacular forests, secluded beaches and some of the most rare and most endangered plants in the world		

2. Using three highlighters or coloured pens:
   - mark in one colour any advantages and disadvantages that affect people (social)
   - in another colour those that affect the economy (e.g. work)
   - in a final colour those that are environmental.

3. **a** In pairs, discuss the following statement:
   'The advantages of the five proposals for the Daintree would be environmental, but the disadvantages would be economic.'

   b Using examples, write about 200 words to say whether you agree or disagree with this statement.

4. Which option do you prefer – the economic developments from the last Study (pages 74 to 75) or the environmental proposals here? Explain your answer.

5. If you want to find out more about the Daintree look at the following website:

   http://www.austrop.org.au/daintree.htm

HAZARDS

Study 1
Earthquakes, volcanoes and plates

What causes volcanic eruptions, and earthquakes? What are the effects of these hazards on people living in the area? What strategies can we use to predict and plan for these devastating events? That is what this study is all about.

1.1 Where, and why, do earthquakes and volcanic eruptions happen?

In Gujarat State, India, in January 2001, at least 18 000 people were killed with many thousands more buried under rubble following an earthquake. The earthquake had a magnitude of 7.7 on the **Richter scale**. The eventual death toll probably exceeded 30 000 and more than 500 000 people were left homeless.

In Washington State, USA, in February 2001, one woman died of a heart attack and there were about 250 injuries in Seattle as a result of a 6.8 magnitude earthquake. Billions of dollars worth of damage was caused throughout the state of Washington.

In the recent past, similar earthquake disasters have occurred in Turkey (August 1999), Kobe in Japan (1995), and in San Francisco (1989, with a worse quake in 1906). We have also had major volcanic eruptions, like Montserrat (1996) and Pinatubo (1991). To what extent are these chance events? Do they form part of a pattern? Can we do anything about earthquakes or volcanoes?

Figure 1 Aftermath of the Indian earthquake January 2001, both a physical and human disaster

■ Seeing the pattern

Earthquakes and volcanoes present a potential major hazard for those living in areas affected by them. However, large areas of the world are free of these two potential hazards; earthquakes and volcanoes are not evenly spread across the earth's surface. Look at **Figure 2**, which shows the world distribution of major volcanoes and earthquake activity.

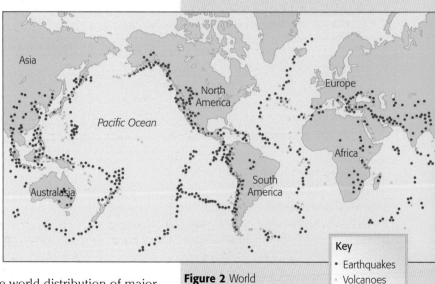

Figure 2 World distribution of major earthquakes and volcanoes

Study 1 Earthquakes, volcanoes and plates

Notice that earthquakes and volcanoes seem to occur close to each other. They can occur on the seabed as well as on land, and they tend to occur along lines. How can we explain this distribution? Look at **Figure 3**, which shows the world distribution of **plates**.

What are plates and how do they move?

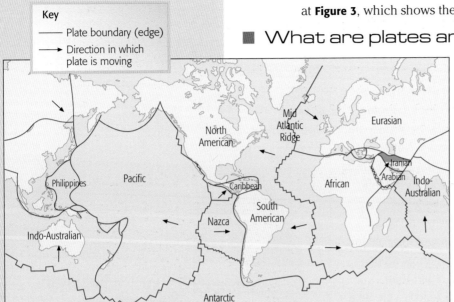

Figure 3 The earth's plates and boundaries

Comparing **Figures 2** and **3** you can see that the distribution of earthquakes and volcanoes closely follows the boundaries between the plates. But what are plates and how do they move? This is all part of a theory known as **Plate Tectonics**.

- Plates are sections of the very thin crust of the earth.
- They 'float' like rafts on the semi-molten material that makes up the earth's mantle, the area immediately below the crust.
- There are two types of plate and their main differences are shown in **Figure 4**. Be careful not to confuse the terms 'oceanic' and 'continental' with ocean and continent. For example, the 'continental plate' may have within it part of an ocean floor.
- It is thought that the development of convection currents within the mantle cause the plates to move, as shown in **Figure 5**. This movement of the plates is very slow, an average of only a few centimetres per year.

	Continental	Oceanic
Age of rock	mainly over 1500 million years	mainly under 200 million years
Weight of rock	light – average density 2.6	heavy – average density 3.0
Thickness	30–70 km on average	6–10 km on average

Figure 4 Plate types

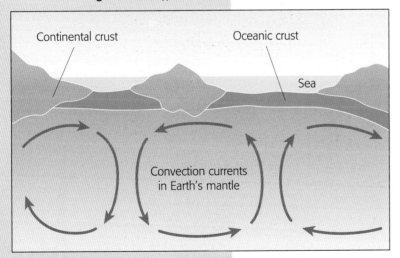

Figure 5 How plates move

The plates are dragged along by these powerful convection currents. Plates may move away from each other or towards each other. They may even slide alongside each other. It is at the boundaries of the plates that virtually all the earthquake and volcanic activity takes place. The central areas of each plate, well away from the boundaries, show very little sign of activity and therefore the hazards from each are very slight.

Over to you

1. South America seems to have a great deal of volcanic and earthquake activity. Explain why this activity is not spread evenly across the continent.
2. Using an atlas, or the internet, find out the names and location of one volcano in each of North America, South America, Europe and Asia.
   a. Identify on which plate boundary they sit.
   b. Research a recent eruption; its date and time, its effects upon people and their environment.
3. Explain why the Pacific Ocean is said to be surrounded by a 'ring of fire'.
4. Why do people who live in the UK have little to worry about in terms of volcanic or earthquake hazards?

1.2 What happens where plates meet?

The boundaries between plates are the places where most earthquake and volcanic action occurs. Four types of plate margin are described here: **constructive**, **destructive**, **collision** and **conservative**.

■ Constructive margins

Constructive margins occur where plates move away from each other, dragged by convection currents. Fresh molten rock or magma rises to the surface and forms new oceanic crust. For example, North and South America are moving slowly away from Eurasia and Africa by about 3 cm a year and the Atlantic Ocean is gradually getting wider. Along this constructive margin, volcanic activity and some earthquake activity can occur. Much of the volcanic activity is in the form of fissure eruptions, where magma pours out through cracks rather than through single vents or holes. This tends to create a spread of lava rather than a cone. The constructive margin in the central Atlantic is known as the Mid-Atlantic Ridge. Iceland sits on this ridge and therefore has a very active volcanic landscape.

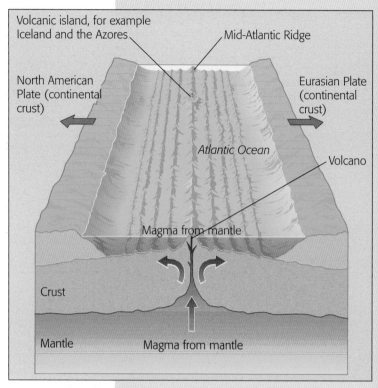

Figure 1 A constructive margin

■ Destructive margins

Destructive margins occur where plates of oceanic crust collide with plates of continental crust. The collision zone is known as the **subduction zone**, and occurs where continental crust overrides oceanic crust. This results in friction and increased pressure as one plate is forced downwards beneath the other. The tension generated can create severe earthquakes. As the descending plate is 'burnt up' by increased heating from the mantle, lighter rocks melt and rise slowly towards the crust, melting more rock as they ascend. This results in volcanic eruptions. Eruptions of this type are rarely of lava (unlike at constructive margins). Instead, they result in huge explosions, which result from trapped water and gases within the destructive zone. Mount St Helens and Montserrat are examples of this type of eruption. They were explosive but generated very little lava.

Increased pressure at destructive margins also creates **fold mountains**. The junction of the Nazca Plate and the South American Plate has created the Andes Fold Mountains with its volcanic cones and its periodic earthquake activity.

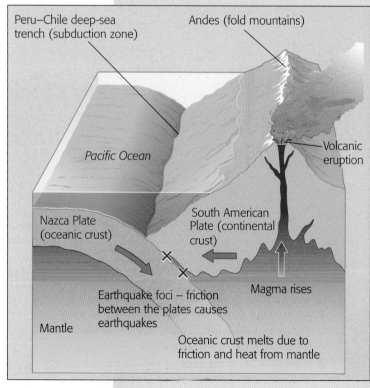

Figure 2 A destructive margin

Study 1 Earthquakes, volcanoes and plates

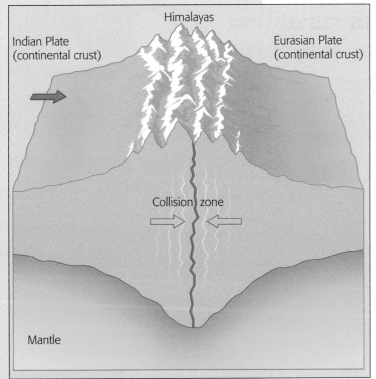

Figure 3 A collision margin

Collision margins

Collision margins occur where plates of continental crust move towards each other. Here neither plate can sink and be destroyed. This means that the edges of the two plates buckle upwards to form large mountainous areas. A prime example is the collision of the Indian and Eurasian Plates, the mountain range formed being the Himalayas, which are still slowly gaining in height. The collision of plates also produces major earthquakes.

Conservative margins

Conservative margins occur where two plates move alongside each other. This margin does not produce volcanic activity or new landforms but can give rise to major earthquakes. Probably the most famous example is found in California, where the San Andreas Fault indicates the margins of the North American Plate and the Pacific Plate. Again, the movement of the plates is very slow but also very uneven. Sudden movements take place following the build-up of pressure along the margin. This sudden 'jerk' as the plates move creates major earthquakes.

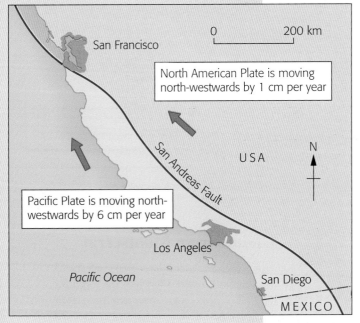

Figure 4 A conservative margin

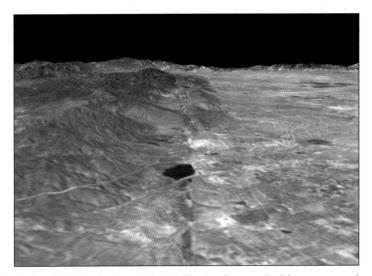

Figure 5 The San Andreas fault in California, photographed from space and enhanced digitally

Over to you

1 a Explain what happens along the Mid-Atlantic Ridge.

  b The Mid-Atlantic Ridge is opening up and 'spreading' at the rate of about 3 cm a year. On a world map identify how it might look in 150 million years time.

2 Why are earthquakes less likely to occur at constructive margins rather than at destructive margins?

3 Draw a labelled diagram to show the sequence of events leading to destruction of crust along a subduction zone.

4 Why should earthquakes be particularly severe along conservative margins?

81

HAZARDS

1.3 Montserrat, a major volcanic eruption

If you have seen the film 'Dante's Peak', you will remember that the most dramatic part occurs when the two characters have to escape an avalanche of rock and ash that follows the beginning of a volcanic eruption. It's Hollywood, and so the characters make it to safety, but the film is engaging and tense nonetheless. It is easy to watch because we are not involved in the disaster. But imagine what it must be like if you are.

The residents of Montserrat, a small island in the Caribbean, had to face a real disaster. Between 1995 and 1997, the island volcano – Chances Peak – erupted after lying dormant for centuries.

Montserrat Fact file

- Settled for tens of thousands of years
- Known to white settlers after it was discovered by Columbus in 1493
- Colonized by the British in 1632 for its trade potential
- Still a UK colony
- Approximately 100 sq km in area
- Part of the Caribbean arc of islands
- Population in 1995 just over 11 000

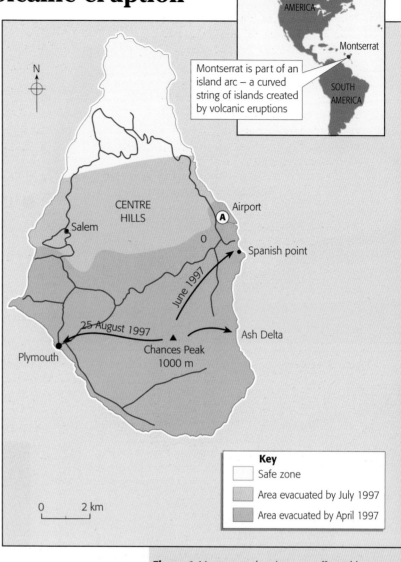

Figure 1 Montserrat showing area affected by the eruption

What happened?

- July 1995 – Chances Peak, a dormant volcano, erupts after 350 years of non-activity.
- August 1995 – two-thirds of the population are evacuated to the north of the island.
- April 1996 – the capital of the island, Plymouth, is evacuated.
- June/July 1997 – **pyroclastic flows** (rivers of hot gas, ash, mud and rock moving at very high speeds at temperatures of about 500°C) cover large areas of the south of the island destroying many buildings by fire. Valleys and forests are covered in ash or lava and 23 people are killed. Two-thirds of the island's housing and three-quarters of the infrastructure is destroyed. Over half the island's population left Montserrat, mainly for the neighbouring island of Antigua, the USA or the UK. Many others were forced into temporary homes.

Update from Montserrat

Seventeen adults and children have been inside the 25 x 20 foot Pilgrim Holiness Church at Davy Hill for two years since they were evacuated from their homes beneath the volcano when it first erupted. At first there were also 40 people camping outside. There is still no gas for cooking, and rubbish is strewn everywhere. Every month food vouchers for £24 are given to adults with £7 for children but this is spent after only two weeks. After that they have to scour the island for food. Life is very bad. There is no privacy and nothing to do. They just wait for help that never seems to come.

Figure 2 Update from Montserrat

Study 1 Earthquakes, volcanoes and plates

Figure 3 Devastation in Montserrat

Plymouth – devastated

PLYMOUTH, the once thriving capital, is coated in thick ash. Face masks are needed to avoid lung damage from inhaling the ash. The noxious smell of sulphur is everywhere. There is a sense that this land will not be farmed again for generations. Only the animals remain.

The villages of Trant, Harris and Streatham, where 23 islanders had died, were completely covered in a light brown carpet of lava and ash. Whole valleys and forests had vanished under thick volcanic mud, the flow stopping only 45 metres short of the single strip runway of Blackburne Airport.

Figure 4 Pyroclastic flow forms a 'delta' as it reaches the Caribbean after the eruption

Figure 5 Plymouth is deserted – covered by up to 3 metres of ash

■ What were the effects?

- Before 1995 – the population had been slowly declining from 14 333 in 1946 to about 11 000 in 1995.
- Each tourist season, mainly villa tourists from North America increased the island's population to between 13 000 and 15 000.
- By November 1997 – the estimated population had dropped to less than 3500, as people left the island.
- By May 1999 – the estimated population had risen again to 4500.
- The population structure had also changed. The proportion under 30 had declined from 52% in 1990/1991 to 34% in November 1997. The average age of the population had also risen, from 28.9 in 1990/1991 to 40 in November 1997.
- Of those remaining on the island, 74% of the population had been forced to relocate.
- A total of 92% of the population had been forced to move, either off the island, or relocated on the island.
- In November 1997, 18% of the islanders remaining were still in shelters and 20% were sharing accommodation as guests or hosts. Higher-income families had tended to move into rented accommodation.

■ Will the island recover?

Scientists believe that Montserrat will eventually recover from the disaster. They use, as evidence, the fact that the neighbouring island of Martinique was devastated by the eruption of Mt Pelée in 1902. Within a few years the devastated areas were being re-populated and the tropical rainforest, which had been destroyed by the eruption, was thriving afresh. The island now has a population of over 300 000.

Once the eruptions on Montserrat stop, scientists will then be able to calculate when it is safe to return to the south of the island. Montserrat has the same tropical environment as Martinique and so the vegetation should thrive in the same way. Already, green shoots are appearing through the ash. In time, the people of Montserrat will be able to build on top of the flows. Some of the volcanic remains could also become tourist attractions as on Tenerife in the Canary Islands.

Over to you

1. What caused the volcano to erupt? Refer back to pages 80–81.
2. a Why are pyroclastic flows so dangerous?
   b Why are pyroclastic flows only found along destructive, and not constructive margins? Refer to pages 80–81 to help you.
3. Draw a table to list the main effects of the volcanic eruption on Montserrat. Subdivide the table into the effects on people, the economy (include farming, communications and tourism), and the environment.
4. What actions could the government have taken to ease the conditions for those people left on the island after the volcano erupted?
5. What evidence do the scientists use to support their view that, given time, the island will recover from this disaster? Does this seem realistic?
6. Which industry on the island could eventually be boosted as a result of the volcanic eruption? Why?

HAZARDS

1.4 Can we predict volcanic eruptions?

Volcanic eruptions can have a devastating impact on people and the environment, as **Figure 1** shows. Being able to predict the eruption, and plan for it, can help to minimize its impact in terms of loss of life, damage to possessions, and so on.

Within the past 25 years, scientists have become increasingly successful in predicting volcanic eruptions. They use several techniques:

- lasers to detect the physical swelling of the volcano. This swelling starts to occur as the magma forces its way up through the vent towards the surface;

- chemical sensors to measure increased sulphur levels. Gases often seep from the surface of the volcano, but the higher the sulphur content within these gases, the nearer the volcano is to the start of an eruption;

- seismometers to detect the large number of miniature earthquakes that can occur as a result of the magma rising up through the vent. These can often be detected a long time before the magma actually reaches the surface, so plenty of warning is available;

- ultrasound, which can monitor low-frequency waves within the magma as the surge of gas and molten rock moves upwards;

- satellite images to record the warming of the ground surface as the magma edges towards the 'breakthrough point'.

Major world eruptions		
1550 BC	Santorini, Greece	destroyed island
AD 79	Vesuvius, Italy	destroyed Pompeii
1669	Etna, Italy	20 000 killed
1792	Mount Unzen, Japan	10 000 killed
1815	Tambora, Indonesia	90 000 killed
1883	Krakatoa, Java	36 380 killed
1902	Mont Pelée, Martinique	26 000 killed
1980	Mount St Helens, USA	66 killed
1984	Nevada del Ruíz, Colombia	24 000 killed
1991	Mount Unzen, Japan	38 killed

Figure 1 Major world eruptions

Figure 2 Monitoring volcanic activity, a sometimes hot and dangerous job. It is only through studying existing hazards that we can make predictions about the future

Study 1 Earthquakes, volcanoes and plates

■ Can prediction help?

Within an 8 km radius of Vesuvius, overlooking Naples in Italy, live nearly 1 million people. A sudden eruption without warning could lead to the death of most of them, especially if a pyroclastic flow was involved. Plans exist to move all the people away from the danger zone within 7 days. But, will there be enough warning?

The eruption of Mount Pinatubo in the Philippines in 1991 does give some encouragement. The first sign of activity on this 500-year-dormant volcano occurred in April 1991. After close monitoring by scientists, a warning to evacuate the area was given and over 75 000 people were moved away. The eruption took place on 15 June with pyroclastic flows and mud flows destroying empty farms and villages and yet only 76 people were killed instead of thousands.

■ What about planning?

Prediction alone is not enough. People, often in large numbers, will need to be evacuated from the danger zone. This needs a high level of detailed planning, as well as resources, and a well developed infrastructure. What exactly do you need?

- for evacuation you need transport and good roads
- for those evacuated, temporary homes
- these temporary homes will need water supply, sewerage, and so on
- those evacuated will need food supplies and medical facilities
- face masks may be needed to prevent the inhaling of ash
- government agencies, such as police, to organize the evacuation
- money to cover the cost which could run into millions of pounds.

As a result, the poorer the country, the less likely it is to have the money, or the facilities, to organize the evacuation of large numbers of people. Also, they will almost certainly lack the equipment for monitoring the eruption and, therefore, be less likely to be able to predict it.

Figure 3 Mount Pinatubo erupts again in 1996

Over to you

1

a Using **Figure 1**, what evidence is there that MEDCs are more able to cope with volcanic hazards than LEDCs?

b What evidence is there to suggest that modern prediction can save lives?

c Why are volcanoes a major hazard for large numbers of people when prediction monitoring is possible?

2 Research to find out why so many people choose to live on the slopes, or very close to volcanoes when they can be such a threat to life.

3 You are responsible for drawing up plans for the possible evacuation of 150 000 people from alongside volcanoes, one located in an MEDC and the other in an LEDC. List any problems you are likely to face in each case. To what extent can these problems be overcome?

HAZARDS

1.5 Turkey, August 1999, a major earthquake

In August 1999 a devastating earthquake hit Turkey. The village of Derince, near Izmit, was the **epicentre**. The earthquake killed over 10 000 people, left up to 35 000 missing and destroyed over 42 000 buildings in Izmit and the surrounding villages.

Turkey earthquake Fact file

- earthquake measured 7.4–7.5 on the Richter Scale
- centred on the city of Izmit, east of Istanbul
- occurred when two blocks of the earth's crust snapped apart
- the movement took place along the North Anatolian Fault (this is similar to the San Andreas Fault in Southern California)
- in places the ground shifted as much as 2.5 metres
- many thousands of people were killed (most buried under collapsed buildings), missing or injured
- caused tidal wave in Izmit Bay

The Richter Scale

The **Richter Scale** is a scale from 0–8 used for measuring the magnitude of earthquakes based on recordings from a seismograph 100 km from the epicentre of the earthquake. The larger the number, the larger the disturbance, 7 being a major earthquake. The epicentre is the point on the earth's surface immediately above the **focus** of the earthquake.

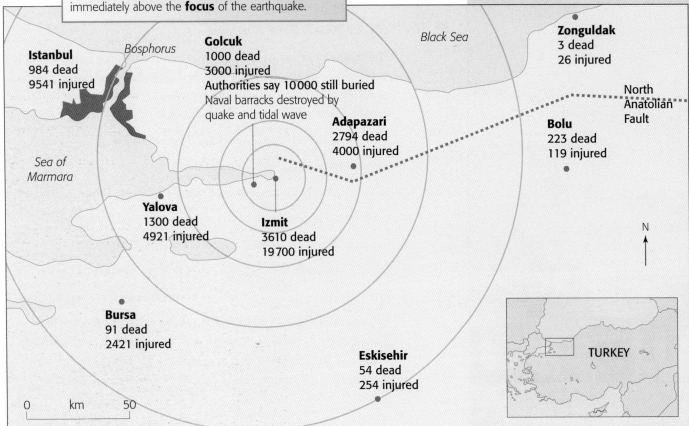

Figure 1 Turkey – destruction and death, earthquake 1999

Figure 2 Reporting the effects of the earthquake

THIRTY-SIX hours after the earthquake reduced Izmit, a city of 1 million people to not much more than a pile of rubble, people still dig desperately, often with just their bare hands, trying to find their missing loved ones under the collapsed buildings. Power lines are strung across the streets. There is no electricity, water or food.

Even after all this time people are still being found alive. Mrs Karakol was pulled from the wreckage of her appartment block by her cousin, who had never given up hope of finding her alive, after hours of burrowing with just bare hands. Mrs Karakol had survived, but had watched helplessly as her husband and four-year-old daughter had died beside her under six floors of collapsed concrete.

Study 1 Earthquakes, volcanoes and plates

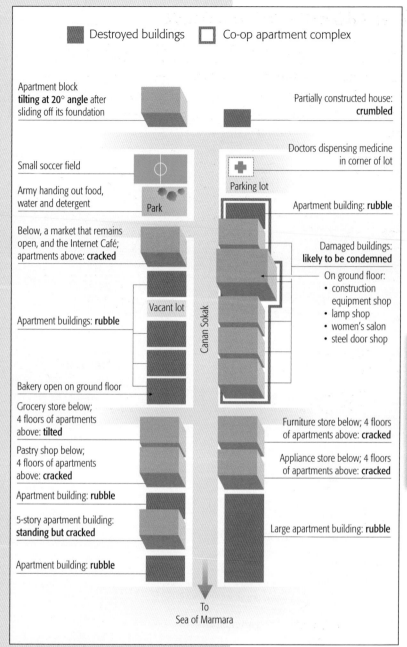

Figure 3 Destruction in Derince, 9 of the 21 houses that stood on Canan Sokak, a street in the Turkish town of Derince, were reduced to rubble by the earthquake

In Turkey, the question most people asked was, 'why should the modern buildings collapse?'

Many of the modern apartment blocks were very cheaply constructed. Beach sand was often used for the concrete in order to save money in construction, and increase the profits of the developer. Also, because buildings were not supervised when under construction, insufficient steel may well have been used so that they could not resist the shaking of the ground when the earthquake happened. In many of the towns and villages it is the newer buildings that have collapsed and not the old.

Figure 4 The effects of the earthquake in Derince

TO WALK down the streets of this town is unbelievable. Many buildings are totally demolished while those that do still stand are uninhabitable, they are so badly damaged. It is not only the old buildings that have collapsed. A six-storey apartment block is now only a collection of twisted metal rods and small blocks of concrete. In this apartment block alone, many died, being buried alive.

Figure 5 Whilst the older buildings remain standing, many newer ones are reduced to rubble, as in Golcuk, close to Derince

Over to you

1 What caused the Turkish earthquake in 1999?
2 What was the main cause of the loss of life? Explain why this occurred.
3 Many of the Turkish towns affected had many new buildings. Why did the earthquake cause them to collapse?
4 What recommendations would you make, knowing that this happened?

HAZARDS

1.6 Can we predict earthquakes?

Whereas volcanic activity is often the end result of a slow build-up of magma pressure within the earth's crust, an earthquake is a sudden, violent, event. Although the pressure gradually builds as two plates temporarily 'lock together', its release is not a slow or steady movement but rather a sudden lurch forward.

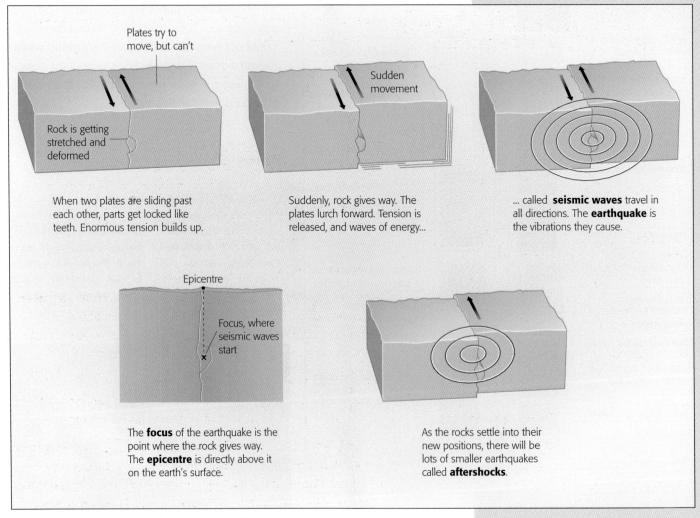

Figure 1 What causes earthquakes?

Because earthquakes happen suddenly, prediction is very difficult. However, techniques are being developed:

- the use of laser beams across a major fault line such as the San Andreas Fault in California where movements of just 1 mm can be detected
- monitoring an increase in the escape of radon gas, which may suggest the approach of an earthquake
- checking water levels in wells; a fall might be as a result of the seepage of water into small tension cracks that are developing
- detecting small fore shocks by seismographs; these may act as a forerunner to a major earthquake.

Study 1 Earthquakes, volcanoes and plates

Figure 2 Modern building techniques can help avoid widespread destruction. Here builders are reinforcing the basement of San Francisco's City Hall

Although we may be able to detect that an earthquake is likely to happen, the precise moment of the earthquake will still be unknown. For example, a major earthquake is thought to be due on the San Andreas Fault in Southern California. But how long before it happens? There was an earthquake in Southern California in 1994 when 57 people died, but this was not considered to be the major earthquake expected at some time in the future. Even that one injured 1500 (in addition to the 57 deaths) and 12 500 buildings were damaged. The actual monetary cost of the damage was in excess of £20 billion.

■ How can we reduce the impact of earthquakes?

In the long term

Modern construction techniques can enable buildings and other structures like bridges to withstand the shaking ground surface caused by an earthquake. Even many older structures can be strengthened so that they will not collapse. This is important, as most deaths associated with earthquakes are due to the collapse of buildings and other structures.

In the short term

A master plan for rescue, restoring essential services, arranging for temporary evacuation, and so on is critical. All the emergency services need to be fully briefed and trained with the appropriate transport available, preferably including helicopters.

People should be educated so that they know what to do in the event of an earthquake. Everyone should have an emergency kit, such as a torch and food, so that they can survive if trapped.

The difficulty of prediction means that, for the most part, long-term preparation is the most effective way of reducing the potential death toll. Constructing buildings or strengthening buildings so that they can withstand earthquake shocks is vital. However, this is expensive and most LEDCs cannot afford to do it, so that when an earthquake does occur, the loss of life is still very great.

Planning for earthquakes

- Architects and builders must follow rules for safe buildings
- All bridges must be strengthened
- Power stations must be built as far as possible from the fault
- Prepare an emergency plan for helping people who get injured
- Establish an earthquake centre where people can phone for advice
- Tell people what to pack in their emergency kit. Things like food and torches…
- Every 3 months schools must practise what to do in an emergency
- Prepare an emergency plan for warning everyone about the earthquake

Figure 3 Planning for earthquakes leaflet

Over to you

1. Why do you think that large numbers of people continue to move into areas such as Southern California, given the likelihood of potential earthquakes?
2. You are responsible for the preparations in the event of an earthquake hitting a densely populated area in an MEDC prone to earthquakes. Describe how you would prepare for it and then explain how your preparations for a similar event in an LEDC would be different.
3. Research to find out how Japan copes with being in an area of the world liable to experience major earthquake activity. Look at the kind of precautions it takes and consider how effective they are in the light of the most recent earthquake.

HAZARDS

Study 2
Flooding

After the wettest 12 months since records began in 1766, it was no wonder that the UK suffered so much flooding in the autumn and winter of 2000/2001. But the flood risk can be managed to some extent, and there are many reasons why people continue to live in areas at risk from flooding.

2.1 What causes flooding?

Figure 1 shows part of York badly affected by flooding in November 2000 and again in February 2001. This was a common scene repeated across the UK during that winter in areas at risk from flooding. **Figure 2** shows the most vulnerable areas in Britain, and government figures showed that more than one million homes were at risk of flooding.

Not all heavy rainfall or melting snow causes rivers to flood. Why does flooding only occur now and again? Basically, most of the rain falling on to the ground surface works its way into rivers, which carry the water away to the sea. The rest will have been **intercepted** by trees and other vegetation and returned to the atmosphere by **evaporation**.

Figure 1 York – one of the worst affected areas during the floods in November 2000

■ Physical factors

To avoid the likelihood of a flood, much of the rain falling on to the ground needs to be able to **infiltrate** the soil and then gradually make its way to the river through the soil and underlying rock, a process known as **throughflow**. The flood risk will be increased if this infiltration process is interrupted. How can this happen?

- After a long hot dry spell of weather, the soil surface can become baked hard which means that rain-water finds it difficult to soak in. Therefore, much of the rain now has to flow over the surface instead with a rapid **runoff** to the river.
- After a long wet period, the soil might have become so saturated that no more rain will be able to penetrate. This means that again the rain-water will have to run over the surface instead.

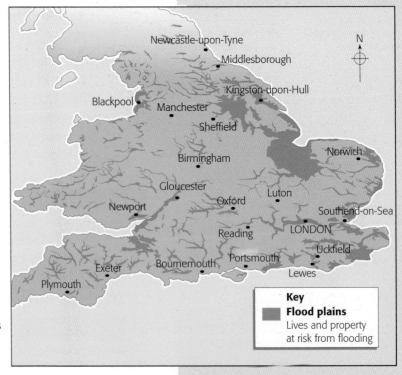

Figure 2 The most vulnerable areas in Britain

Key
Flood plains
Lives and property at risk from flooding

Study 2 Flooding

- The underlying rock may be fairly close to the surface and this may be **impermeable** so that the water cannot penetrate. This will mean that the soil above will quickly become saturated and any surplus water will have to run over the ground surface, so reaching the river quickly.

- Also, in any river valley which has steep slopes, water will tend to move faster downslope than in those where the slopes are gentle.

Human factors

It seems that the risk of flooding is increasing, and that more and more money is needed for flood defence schemes. When flooding does occur, the effects and damage are getting worse. Are human factors contributing to this?

- If trees are cut down this can cause rain to run over the ground surface more quickly and to a greater extent. The removal of trees reduces the infiltration process and increases runoff. However, not all river valleys are surrounded by cleared areas of forest so there must be another possible reason for the increased flood risk.

- The most likely factor is the replacement of open farmland or woodland with buildings, roads and other impermeable surfaces. A field allows water to infiltrate, a concrete or tarmac surface will not. Most river basins in the more populated areas have seen a huge reduction in infiltration as urban areas continue to spread.

- There has also been an increasing tendency to build on river **flood plains**. Many of these rather wet flood plain areas, close to rivers, were left undeveloped in the past, partly because of the risk of flooding but more likely because of the problem and cost of drainage. Today, with increasing pressure on space for building, and improved drainage techniques, these level flood plain areas seem ideal sites for new industrial areas, retail parks and housing. Consequently the river has often lost its floodable land, where surplus water could spread without any real damage. Instead, it will flood buildings, causing great damage. There is even the risk of loss of life.

Figures 3 and **4** Excess water on the floodplain – where human development is sparse, and in a built-up area

Over to you

1. Explain how the type of rock found in a river basin can influence the risk of flooding by the river.
2. How can the clearance of woodland in the river basin contribute to an increased flood risk?
3. Explain how the spread of urban areas onto the flood plain is increasing the risk of flooding.
4. Why is there so much more building on flood plains now than there used to be?

HAZARDS

2.2 Flooding in North Yorkshire, 1999

The skies opened, the rains came and for hundreds of residents in the valley of the River Derwent in North Yorkshire, it was the start of a nightmare. It was the beginning of the worst flooding in that part of Yorkshire for over 70 years. The River Derwent, which drains from the high ground of the North York Moors, burst its banks to leave some streets and properties under 1.5 metres of water. Whole areas of the valley were disrupted and damaged by the devastation. There had been periods of heavy rain before which had not led to serious flooding, so why this time were things going to become so bad?

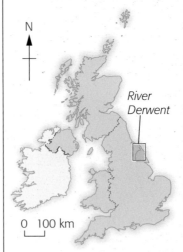

Figure 1 Location of the Derwent river basin

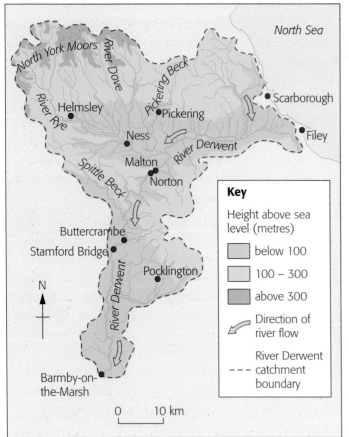

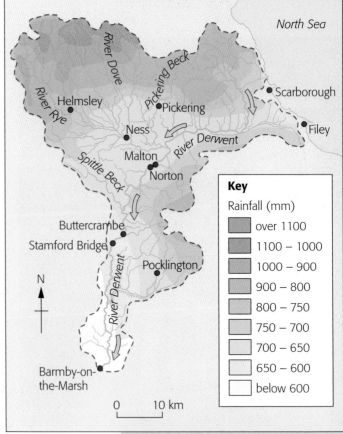

Figures 2 and **3** The Derwent river basin – relief (left) and rainfall

As can be seen on **Figure 2**, the River Derwent rises on the high ground of North York Moors very close to the North Sea coast. It flows southwards for approximately 20 km and then turns westwards. Eventually it turns southwards, being joined by a large number of tributary streams, which also mainly drain from the North York Moors.

Study 2 Flooding

Figure 4 When the ground is saturated, water cannot infiltrate, and will flow over the surface

What caused the flood?

- Near stationary low pressure system over the North Sea which caused prolonged heavy rain over the North York Moors, the area where the River Derwent and its tributaries rise – 150 mm of rain fell in 3 days.
- Melting snow on the North York Moors added to the surface runoff.
- Saturated ground further down the valley – no infiltration was possible and so surface runoff was greatly increased.

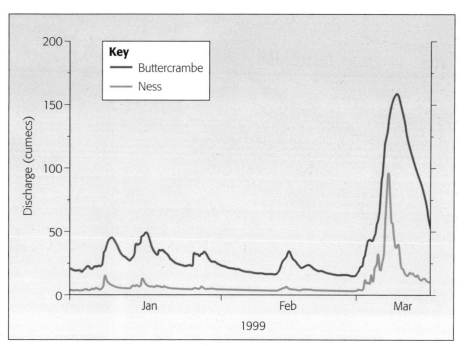

Figure 5 Two **hydrographs** are shown for the Derwent river basin. One for the Ness recording station and the other for the Buttercrambe recording station. The discharge of the river has been recorded over the period January to March 1999. The discharge has been recorded in cubic metres per second

1998 had generally been a wet year over much of the UK, but the rainfall in February 1999 within the basin of the River Derwent had not been exceptional (parts of the North York Moors receiving a total of around 45 mm of rain for the month against the long-term average of about 55 mm). Nevertheless, the ground was close to saturation so that any further rain would have an immediate impact on the River Derwent and its tributaries. Heavy rainfall was to follow. Parts of the North York Moors received over 250 mm of rain between 28 February and 11 March 1999, whereas the long-term average for the whole of March was only 75 mm.

HAZARDS

2.3 What were the effects of the flood?

Waters continue to rise – rivers on 'Red Alert'!

March 1999

Parts of North Yorkshire were facing their worst floods for 70 years as the River Derwent burst its banks. The first town to be flooded was Pickering, although the flood level is falling now as the flood surge moves further down river. Already some homes are flooded in Malton and more homes are being evacuated in preparation for the peak which is likely to arrive tomorrow. In Malton and the surrounding area, some roads and houses are already under three feet of water. At Stamford Bridge, where the peak is still two days away, Main Street is already like a river. In one of the pubs the water is waist deep in the bar.

Figure 1 Reporting the floods (1)

Figure 2 The centre of Stamford Bridge is cut off by flood water

Figure 3 Boats had to be used to rescue people in the flooded areas

Study 2 Flooding

March 1999

Floods bring chaos to North Yorkshire

People living in a cul-de-sac in Pickering were stranded in a 'lake' of raw sewage! It just bubbled out of the drains! The residents of 40 houses were trapped in their homes until contractors could start to pump it away.

Commuters in Malton were left stranded as the rail link between York and Scarborough was cut. Although a bus service had replaced the rail service, the bus could not get into Malton.

Canoes and boats were the only way to get about in Stamford Bridge and many other towns and villages. Many minor roads were blocked by flood waters and the A169, Malton to Pickering road was completely cut, as was the A166 to Stamford Bridge.

Figure 4 Reporting the floods (2)

Figure 5 The flooding caused more than just inconvenience; considerable damage resulted from the water entering buildings

Derwent river basin Fact file

Area – 2057 sq km

Main towns – Pickering, Norton, Pocklington, Malton, Helmsley, and Stamford Bridge, each with a population of less than 6000.

Population of river basin – approximately 100 000, although in summer, tourists may increase this considerably.

Agriculture – in the upper part of the river system it is mainly grazing, but in the middle and lower sections, arable land becomes predominant.

Transport – the A64(T) from Leeds to Scarborough runs south-west to north-east through the central part of the drainage basin with the A170 running east-west across the northern part. A whole network of B roads and minor roads criss-cross the entire river basin, apart from the North Yorks Moors from which most of the tributaries drain.

Over to you
Activities relate to pages 92–95

1. What were the main causes of the flooding in North Yorkshire?
2. Describe the changes in the river flow between January and March as shown on the hydrograph in **Figure 5** on page 93.
3. Explain why places further downstream are likely to receive more warning of a possible flood than places further upstream.
4. Explain why the extent of the flooding is likely to get worse as you move downstream.
5. Study all the photographs on pages 94–95 carefully. Make a list of all the possible effects of the flood for people living in the Derwent valley. Divide your list into two parts, one listing the short-term effects and the other listing the long-term effects.
6. Suggest some of the measures that could be taken to reduce the possible impact of any future flood in this area. Consider both short-term and long-term measures.

HAZARDS

2.4 Flooding in Mozambique, 2000

Flood horror!

As the helicopter got closer to the tree, we could see flashes of colour – the headscarves, clothes and bundles of people – people who had spent four days and four nights desperately clinging with their children to the very top branches of the tree. During the day the temperature had risen to 104°F while at night, the mosquitoes returned. All the time, they faced the constant fear of the ever rising flood waters.

Slowly and surely all were winched into the helicopter. Salima was one of those rescued, a six month old baby on her back and a chicken tucked under one arm, her most valuable possession! Salima had been on her way home from church when the flood struck. The water had come so quickly and with such force that all she could do was to grab the chicken and climb the tree.

Figure 1 The human drama brought to the attention of the world

Scale of the disaster becomes clear – over 1 million lose their homes, thousands drown

More than one million people have lost their homes and many thousands have been drowned – the precise number will probably never be known. Yet it is only when you hear people's stories that the real scale of the disaster becomes clear. At one refugee camp Violeta told us how she had watched her 6 year old daughter drown, unable to do anything because of her young baby on her back.

These people had so little – the average income is less than £65 a year. The one cow, the one goat, the few chickens – their only possessions of value had been built up over many years or even generations. Now all this has gone.

Figure 2 The effect of the flooding in a country that is already very poor

Study 2 Flooding

Rescue missions go on!

After rescuing 435 people in eight hours, our pilot was ordered back to base. However, just as we were about to turn, we spotted a rooftop crowded with people. They waved at us frantically, but we just had to leave them behind, to wait through another night to be rescued hopefully tomorrow! I shall never forget the look on their faces as we turned away.

Figure 3 The impact of a serious disaster goes further than just the immediate victims

Figure 4 Loss of infrastructure such as roads and bridges causes more problems

Refugee camp horror!

At one of the refugee camps where more people are arriving all the time, several thousand people wait. They have just one tent, no sanitation, little food and just one small hut with a red cross on it. It has no drugs, only one small first-aid kit. How do you get food and clean water to them when all the roads and bridges have been washed away?

Mozambique Fact file

- **Area** – 799 380 sq km
- **Capital** – Maputo (estimated population approximately 2 000 000).
- **Population of Mozambique** – 21 500 000 (estimated) spread over a largely rural and almost totally undeveloped country.
- **Agriculture** – dominated by peasant growers who produce the bulk of the staple crops – cassava and maize. Mozambique is desperately poor, even by African standards. Its main export is prawns. Otherwise its earnings are agriculture based, mainly from nuts, cotton and sugar.
- **Transport** – a single highway connects Maputo to the north.

HAZARDS

2.5 What were the causes of the flood?

The photos and newspaper articles on pages 96 and 97 show the devastating effects of the floods in Mozambique early in 2000. **Figure 1** on this page shows the extent of the areas affected by flooding. So what caused the flooding?

Sequence of events leading to the flood

January 4th 2000 – A three-day storm struck Mozambique and the north-eastern part of South Africa. More than 350 mm of rain fell in one day.

January/February – Almost continuous heavy rain for more than 5 weeks. In February, the rainfall total was 1163 mm compared to the average of 177 mm.

February 22nd – Cyclone Eline struck Mozambique, bringing more torrential rain.

Conservationists blame South Africa!

Their reasons are:

- The destruction of grasslands on the High Veld (high plateaus which run through the central areas of southern Africa). These grasslands have always acted as a sponge, soaking up much of the rain-water and then releasing it slowly into the river system.

- The destruction of the natural wetlands alongside the river system. These areas were the safety valves for rivers in flood. The marshy areas and level flood plains alongside the rivers could temporarily absorb much of the flood water whereas now the flood water has nowhere to go.

- The huge growth of urban areas around the Johannesburg and Pretoria region. This has created a vast area of impermeable paved and concrete surfaces which drastically reduce infiltration and encourage rapid run-off to the rivers.

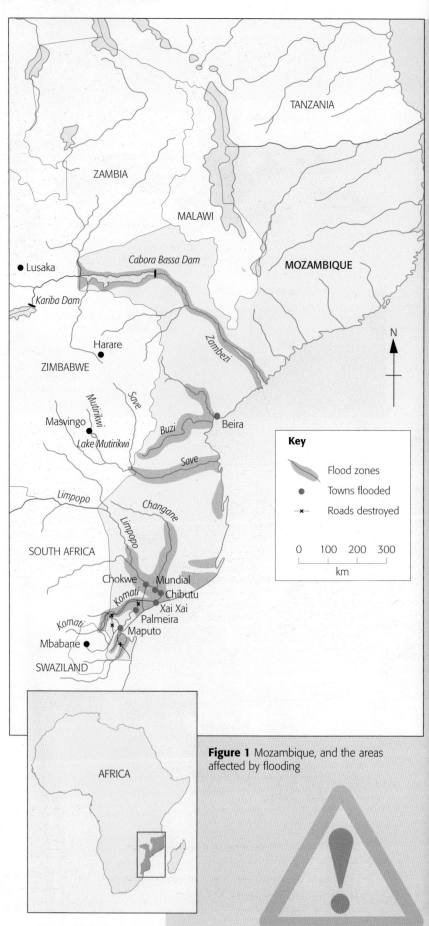

Figure 1 Mozambique, and the areas affected by flooding

Study 2 Flooding

Figure 2 The natural grasslands of the High Veld absorb rainfall by the process of infiltration

Figure 3 The growth of urban areas, such as Johannesburg in South Africa, increase runoff and the possibility of flooding. The extra problem is that rapid run-off in city areas can damage and destroy poorly-built shanty town areas

■ The main cause? — rainfall

In spite of what conservationists claim, there is no doubt that the main cause was the record-breaking rainfall.

Ironically, southern African countries had just set up an EU-funded project to place high-tech equipment in the region to monitor rainfall and river levels. If this had been fully operational, the people would have received warning and been able to get above the flood. In Bangladesh, where there are frequent floods, the development of such an early warning system, together with cyclone shelters and the construction of artificial hills, has greatly reduced the death tolls.

■ Response to flooding

The EU-funded project, mentioned above, to install high-tech equipment in the region to monitor rainfall and river levels was set up in 1999. However, the flooding was so severe that many sensors were washed away or were submerged. The new system could not have prevented the flooding. But had it been operating properly, it could have warned the people to get above the floodwaters.

Over to you

1 What was the main cause of the Mozambique flooding in 2000?
2 What other factors could have contributed to the flooding being so severe?
3 Suggest some of the measures that could be taken to prevent a similar disaster in the future.
4 Although the technology and expertise to monitor and plan for such hazards is available on a world scale, why do so many LEDC countries like Mozambique find it so difficult to take advantage of?
5 Suggest reasons why people still tend to live in areas that could be liable to serious flooding.

HAZARDS

2.6 Can flooding be controlled?

When flooding does take place, a whole range of short-term help is normally available. In MEDCs police and fire services help with evacuation; local authorities provide temporary shelter and voluntary organizations help with food, drink, bedding, and other essentials. When serious flooding affects very poor LEDC countries like Mozambique, international help is usually forthcoming. However, in both MEDCs and LEDCs the longer-term cost is considerable, with damage to property and possessions, and loss of livelihoods, as shown in **Figures 1** and **2**.

Consequently there is a need to try to create long-term solutions to prevent further flooding taking place.

■ What can be done

a Build dams and storage reservoirs in the upper part of the river system – the rain-water falling onto the higher ground can be stored and then slowly released downstream at a rate that the river channel can cope with.

b Create temporary storage areas – land that water can be pumped onto so that the floodwater does not reach the more populated areas. This collected water can then be gradually pumped back into the river once the flood threat has passed.

c Improve the channel of the river by making it deeper or straighter – the river can cope with the extra volume more efficiently. A deeper, wider channel can hold more water, and a straight channel rather than a winding one will speed the flow of water towards the sea.

d Arrange for flood banks or flood walls to be built alongside the river – this creates extra capacity in the river so that the extra water generated in the flood stays within the river channel.

e Preserve part of the flood plain alongside the river just as an area of floodable land – such land could be used for grazing animals under normal conditions, but allowed to flood if needed. These 'washlands' create the extra space needed for the flood waters and enable the rivers to flood, which is an entirely natural process, without causing costly damage or possible loss of life.

f Reduce the impact of possible flooding – not allowing the construction of houses or factories, or even roads, close to the river, on the most vulnerable parts of the flood plain. There is no doubt that building on flood plains makes the problem worse.

Figures 1 and **2** Damage caused by flooding in North Yorkshire (above) and Mozambique. The countries are very different but the effects are similar

Study 2 Flooding

Flood prevention on the River Derwent

- Most of the floodable land along the Derwent and its tributaries is still rural with relatively few settlements close to the river – hence flood prevention has not been considered so vital.
- Most of the main rivers in the Derwent system have long-standing flood banks which provide protection for the adjacent farmland.
- Extensive flood plains exist to act as temporary storage for flood waters.
- Washlands have been created further downstream between the embankments and river, again to temporarily hold any floodwater.
- Urban developments on the flood plain have to take into account their impact on surface run-off and the addition to the flood risk.
- Telemetry stations are in place to provide data on rainfall and river levels, so that a forecast can be made of likely flooding. Thus, early warnings can be given to those living in areas at risk of flooding.
- Local emergency services have a strategic plan for possible evacuation for towns such as Malton and Stamford Bridge.
- TV and radio are used to broadcast flood alerts and a Floodcall Telephone Information line is provided. As a last resort, a loudspeaker system would be used.

Because of its rural catchment, there are no new flood prevention measures planned along the Derwent. But there will be continued maintenance and minor improvements to existing systems.

Figure 3 Flood protection on the River Derwent. Long-term remedies may still be costly for LEDCs.

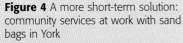

Figure 4 A more short-term solution: community services at work with sand bags in York

Flood prevention in Mozambique

Apart from the EU-funded system that has been put in place, the rivers in Mozambique have no major flood prevention schemes. In some places embanking does exist, but not in a coordinated way. This is mainly because Mozambique is one of the poorest countries in the world, ranked 195th in terms of per-capita GNP. But it's also because serious flooding is not a frequent event, meaning the hazard has not attracted world media attention like it has in Bangladesh. Bangladesh eventually got MEDC financial support for flood prevention measures.

Compared to the small-scale River Derwent system, the river systems in Mozambique are vast, draining across international frontiers. This means river control and flood prevention is much more difficult. Making Mozambique safe from flooding would be very expensive and would also need complex international agreement.

Over to you

1. Suggest why many towns and cities have allowed building on flood plains. Give examples of any close to where you live.
2. Why would building on a flood plain make flooding itself worse?
3. Consider the long-term solutions for controlling flooding given in (a) to (f) on page 100. Take each one in turn and think what the problems would be in trying to put it into practice.
4. Most of the flood prevention measures in the valley of the River Derwent are quite basic and involve relatively low cost. Explain why, for a very poor small country like Mozambique, such measures are still difficult to put in place.

POPULATION DYNAMICS

Study 1
How is population changing?

Population dynamics is all about population change. A country's population is made up of millions of people just like you.

This chapter will look at how, and why, population is changing; how countries respond to changing numbers of people; why different countries have different numbers of people of different ages and how they can plan for, and cope with this.

1.1 The world just keeps on growing

In 1950 there were 562 million people living in China. By 2000 (just 50 years later) there were twice as many. The population had more than doubled to 1261 million. China was afraid that there would not be enough food, water and other resources for all its population. (See pages 104–105 for a case study of population change in a Chinese family.)

The population growth that was happening in China was also happening throughout the world.

Look at **Figure 1**. It shows the growth of world population since 10 000BC. You can see how the population has grown since then and how many more people there are now. You can also see a projection for the future.

■ How does population grow?

Figure 2 shows how the world's population grows. It shows the **process** of change. Change depends on the number of births and the number of deaths. The difference between the number of births and the number of deaths is called the **natural increase** in population.

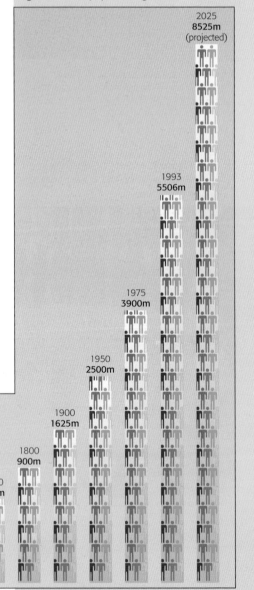

Figure 1 World population growth

Study 1 How is population changing?

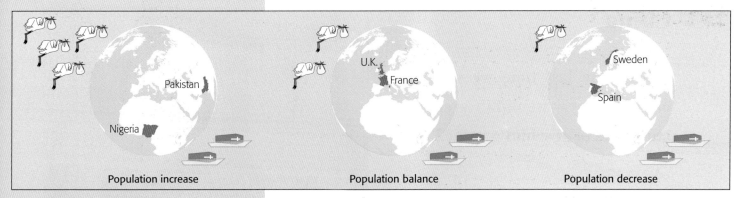

Figure 2 What do these diagrams tell you about population change in these countries?

Three examples of population change are shown in **Figure 2**. In each diagram, two countries are given as examples of where this type of change is happening.

How fast is population growing?

Population - key words

When you are learning about population it helps if you know and understand some important words and terms.

Countries with a large number of people may have more *actual* births and *actual* deaths than those countries with a very small number of people, *but* they may not increase so rapidly. We say that they have a slower **growth rate**.

To work out the **growth rate** of a population we look at the **birth rate** and **death rate**.

Birth rate (B.R.) = number of live births per thousand population per year.

Death rate (D.R.) = number of deaths per thousand population per year.

B.R. – D.R. = rate of natural increase (N.I.) = growth rate. It may be a positive or a negative number, so the population may be growing, or in fact declining (getting smaller).

Some populations also change because of migration. We'll look at that on pages 114–117.

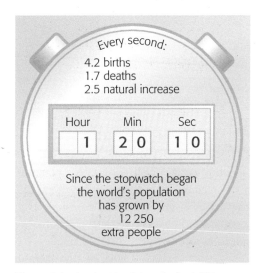

Figure 3 In the time it takes to look at this stopwatch, even more babies have been born!

Figure 3 shows how quickly the world's population grew in 2001. It tells you how many babies are born in the whole world. Every second there are on average more than 4 babies being born. But it also tells you how many people are dying as well. The natural increase shows how many more people there are every second. If you read the stopwatch it will tell you how much the world's population has grown since the watch was set.

Over to you

1 Look at the graph in **Figure 1** that shows the world population growth from 10 000BC to AD2025.

  a Copy and complete the following sentences:
  Between 10 000BC and 4000BC the world population increased from _____ to _____. It more than doubled but it took _____ years.

  b Now answer these questions:
  i) In AD500 the population was 190 million. By what date had it more than doubled?
  ii) How many years had this taken?
  iii) How long did it take to more than double from 1625 million to 3900 million?
  iv) How long is it predicted to take to more than double from 3900 million to 8000 million?
  v) This speed of change is known as **exponential** growth (a maths word). By looking at the doubling times, explain what you think 'exponential' means. You can check in the glossary!

2 Use **Figure 2** to explain why the world's population changes. It may increase, stay the same or decrease. Describe what each diagram shows.

3 Look again at the stopwatch (**Figure 3**). Work out how many more people there would be in the world if the stopwatch had timed only 10 seconds. (Hint: Natural increase × number of seconds.)

Choose some other lengths of time and calculate the extra people.

103

POPULATION DYNAMICS

1.2 China

> 'That's the last time we saw our little boy – we didn't even have time to give him a name.'

This year 13 million Chinese babies will be born. China has difficulty providing food, water, jobs, and facilities for all its people. Laws have been passed to stop people having so many children.

This is the story of the killing of Huang Quisheng's son, a boy too young to have a name.

Figure 1 shows a Chinese family who have kept to the laws.

In **Figure 3** you will read a newspaper report about a family who didn't.

Figure 1 The child in this typical Chinese family will probably have no brothers or sisters … and very few cousins.

China factfile

Facts:	25% of world's population
	7% of world's arable land
	8% of world's water supply
Laws:	1979 – 1 child per family
	1980s – parents in rural areas may have 2 children if first is a girl
	1990s – city parents may buy permit (costing one year's wages) for second child
Punishment:	abortion, mothers held in detention centres, sterilization, fines, eviction from home

Figure 2 China factfile

Ji Quingzhen was eight months pregnant with their fourth child when, at 7pm one evening, three men came to call. Two were called Zhang and another Deng. They said they were family planners who were administering China's population control policy. They made it clear that, as the couple already had three children, this baby would not be allowed to live.

The couple were dragged two miles to the nearest family planning hospital in Caidian, Huang Pi county, in the central province of Hubei, where Ji was forcibly injected with saline solution in an attempt to induce a miscarriage even though the baby was less than a month from its due date.

During an agonizing 15 hours she was taunted by the officials and repeatedly reminded of her unborn child's pending death. At 10am the following morning, doctors and officials gathered around her bed to witness the outcome. Ji finally gave birth – and then came an urgent wail of life. Against the odds, Ji had given birth to a healthy baby boy.

The family planning officials were horrified. Before she could even cradle her newborn, he was snatched away.

'I pleaded with them to allow us to keep our baby, but they just said we had too many children and this would not be tolerated,' Mr Huang told us.

The baby was killed. 'That's the last time we saw our little boy – we didn't even have time to give him a name.'

Andrew Laxton, *The Mail on Sunday*, 3 September, 2000

Figure 3 The story of Huang Quisheng's son

Study 1 How is population changing?

FAMILY PLANNING HOSPITAL, CAIDIAN					
ADMISSION FORM					
Name:	Ji Quingzhen				Address:
Next of Kin:	Huang Quisheng (husband)				Ding Jia Wang
Children: (with ages)	Ting Fang (9)	M/F	F	At school/home	S	Near Caidian, Huang Pi County Hubei Province China
	Zhi Gang (5)	M/F	M	At school/home	H	
	Chuen Fang (3)	M/F	F	At school/home	H	
Occupation:	Mother, housewife and farmworker.				
Partner's occupation:	Farmer growing rice, vegetables, has several chickens and geese.				
Children's activities:	At school; helping on farm/looking after little sister.				
Working hours:	Sunrise to sunset, 7 days a week.				
Accommodation:	Damp, stone-walled house, wet and muddy earth floor, 2 rooms. One chair, one milking stool, one bed, small stove.				
Income:	£25 per year				
Hobbies:	Card games, mahjong, talking with neighbours over glass of rice wine.				

Figure 4 Hospital admission form which shows the Quisheng family's lifestyle: how different is it from yours?

Over to you

1 The news editor of your local radio station has just handed you a pack of information containing **Figures 1 to 4**.

Write a report for your local radio station to explain what happened in China.

People will want to know:
- what the family is like
- where they live
- what they do
- why they wanted another child
- what happened at Caidian Hospital
- why the Chinese officials reacted this way
- what you think about the events.

POPULATION DYNAMICS

1.3 Where in the world is population growing?

There are more and more people in the world every second. In some places the number of people is rising very fast. But in other places the change may be very slow.

As geographers we can look at a world map and see the patterns of change. **Figure 1** shows the countries with the fastest growth rates in *red*, and those with much slower growth rates in *green*.

We usually look at the population growth for each year and then find an average. The map has taken an average for all years from 1980 to 1995. So that big countries and small countries can be compared, we look at the percentage growth, or the growth per thousand people. This map uses percentages.

On pages 112–123 we will look in more detail at the population in Germany (an MEDC) and Malawi (an LEDC). These two countries are marked on the map.

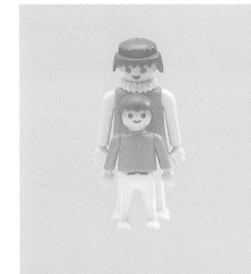

Figure 1 World population growth 1980 to 1995. Where did population grow the fastest?

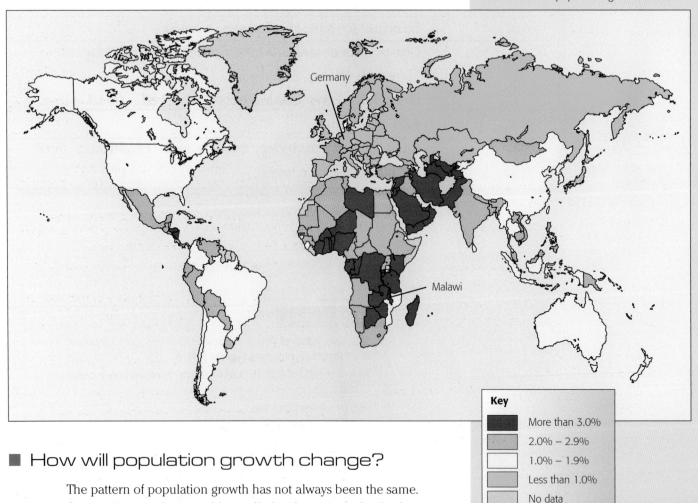

Key
- More than 3.0%
- 2.0% – 2.9%
- 1.0% – 1.9%
- Less than 1.0%
- No data

■ How will population growth change?

The pattern of population growth has not always been the same. Geographers are interested in predicting the patterns for the future, and so they use a **model** or a calculation to work out what is likely to happen.

Study 1 How is population changing?

Figure 2 shows the population sizes of different countries. Starting in 1950 it is possible to see the change to 2000 and then the prediction for 2050. Notice that some countries will probably decrease in population by 2050. Notice also that the countries with the largest population (the ones with the highest rank) in 2000 are not all the largest countries by 2050 because their growth rates change.

Figure 2 Predicted population change 1950–2050

Country	Rank 1950	Population 1950	Country	Rank 2000	Population 2000	Country	Rank 2050	Population 2050
China	1	562 million	China	1	1261 m	India	1	1619 m
India	2	369 million	India	2	1014 m	China	2	1470 m
USA	3	152 million	USA	3	275 m	USA	3	403 m
Russia	4	101 million	Russia	6	146 m	Nigeria	5	303 m
Japan	5	83 million	Pakistan	7	141 m	Pakistan	6	267 m
Germany	7	68 million	Japan	9	126 m	Russia	14	118 m
UK	9	50 million	Nigeria	10	123 m	Japan	16	101 m
France	12	41 million	Germany	12	82 m	Germany	23	79 m
Pakistan	13	39 million	UK	20	59 m	France	29	58 m
Nigeria	15	31 million	France	21	59 m	UK	30	58 m
Spain	17	28 million	Spain	28	39 m	Nepal	32	53 m
Nepal	38	8 million	Nepal	41	24 m	Spain	53	32 m
Sweden	53	7 million	Malawi	70	10 m	Malawi	79	14 m
Malawi	91	2 million	Sweden	82	8 m	Sweden	108	8 m

Key: Countries with an increase in population | Countries with a decrease in population

Over to you

1 Look at the table in **Figure 2**.

  a Describe what happens between 1950 and 2050 to i) China, ii) India and iii) USA.

  b Which of these three is expected to grow fastest in the next 50 years (2000–2050)?

  c What happens to countries like Pakistan and Nigeria between 1950–2000 and 2000–2050?

  d What happens to Nepal and Malawi?

  e What happens to the rank of the UK and France? They can't keep up in the population race. What is happening to the size of their population:

  between 1950–2000?

  between 2000–2050?

  f What is happening in countries like Spain, Russia and Germany?

2 If you want to know more about the population growth of other countries, look at www.census.gov/cgi-bin/ipc/idbrank.pl and www.worldbank.org/depweb/english/modules/social/pgr

3 Look at the map in **Figure 1**. Using your ideas from question 1 and the map, find out if more population growth is happening in MEDCs or LEDCs?

4 Find evidence to support your answer in question 3 from **Figure 2**. Make a list of MEDCs and a list of LEDCs taken from **Figure 2**. Put them in a table like the one below.

  a Write in their actual growth (number of extra people), from 1950–2000.

  b Then look at growth rates (the percentage or proportion of extra people). Percentage growth = number of extra people ÷ number of people before x 100. Put the figure in the % growth column.

  c Add up the totals for actual growth for MEDCs and LEDCs. Did LEDCs or MEDCs have more actual numbers?

  d Did LEDCs or MEDCs have higher percentages?

  e Where is the fastest growth happening? Use evidence from your table to explain your answer.

MEDC	Actual growth	% growth	LEDC	Actual growth	% growth
Country name			Country name		
Country name			Country name		
Country name			Country name		
Totals					

POPULATION DYNAMICS

Study 2
Why is population changing?

What causes different countries' populations to change in size? Why are there more people in some age-groups than in others? What effect do people moving from one place to another (migration) have on population? These are some of the questions that this study will look at.

2.1 Why do some countries' populations grow faster than others?

Pages 106–107 showed that some countries' populations grow in size faster than others. But why does this happen?

It's all to do with the number of children that each woman has – sorry boys! The reason is of course that it's only the women who actually have the baby. If there are extra men around they don't add newcomers to the population.

Take a look at the two couples shown here in **Figure 1**. We have three different possibilities for them. What happens if the women in each generation have 2 babies,

> or 3 babies between them – an average of 1.5 babies each,
>
> or 5 babies between them – an average of 2.5 babies each?

Look at **Figure 1** and see how many people there are in each generation for **a**, **b**, and **c**.

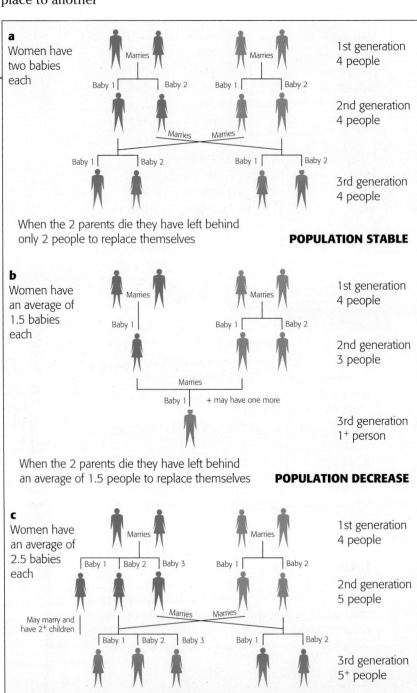

Figure 1 What happens when people have children

108

Study 2 Why is population changing?

Figure 2 How do people decide how many children to have? It's different in MEDCs and LEDCs

If each couple only produces two people then, when the two of them die, they leave behind only the same number (two) in the next generation. They leave their own replacement. This is what happens in **Figure 1a** and the population is stable, neither increasing or decreasing.

In **Figure 1b** they do not replace themselves and the population decreases.

In **Figure 1c** they leave behind far more people in the next generation so the population increases.

■ Population growth in LEDCs and MEDCs

In many parts of the world there are still big differences in the number of children people have, and the reasons why they have them.

In LEDCs many parents want to have several children so that they can help with the work at home and on the land or earn money in the city. In MEDCs where families are, on average, richer and more educated they do not need the children to earn money for the family. They usually have fewer children.

Figure 2 shows these different views of children.

Over to you

1. Look again at **Figure 1**. Now draw a picture to show what happens if each couple only has one child.
2. China had a one-child policy which was the law. There they wanted to stop the population growing and they wanted to have boys. What do you think would happen if they killed all the baby girls and only allowed the boys to live?
3. Look at **Figure 2** and explain why you think that families in LEDCs often want more children than do families in MEDCs.
4. You have been sending e-mails to Abi, a friend in Malawi. Malawi is a small LEDC country in Africa. You told her that if all girls only had two babies, then the world population would stabilize (not get any bigger).

This is her reply:

```
From:     abi@malawinet.com
To:       gcsefriend@school.org
Sent:     09.45
Subject:  Babies!
  Hiya!
  Me again. You've got me confused.
  O.K. I'll only have 2 babies BUT
  how does that help? There'll still
  be 2 more people in the world.
  Love, Abi.
```

From what you have learnt, write your reply.

109

POPULATION DYNAMICS

2.2 What else causes different patterns of population growth?

So lots of births cause lots of population growth. Is that it? Well, not quite. It's not as simple as that!

The two pictures in **Figures 1** and **2** are taken from Victoria Brown's family picture album. They show all of the Brown family in 1904 and then all Victoria Brown's relations (she's now a great-grandmother) in 2000.

In **Figure 1**, Victoria is the baby and, there were more children than adults. In 1904 many children and adults died at a much younger age than they do now. Victoria only knew one of her grandparents. The others had already died before she was born.

In **Figure 2** the little baby still has three grandparents who are alive. He also has two great-grandparents including Victoria (seated). In this picture there are far more adults than children.

Population growth is not just about births. It also depends on how long people live, **life expectancy**. If people live for longer then there are more people who are still alive on the earth and the world population grows. If people live longer the death rate is lower.

Figure 1 Victoria Brown's family in 1904

Figure 2 The family in 2000: what had changed?

Study 2 Why is population changing?

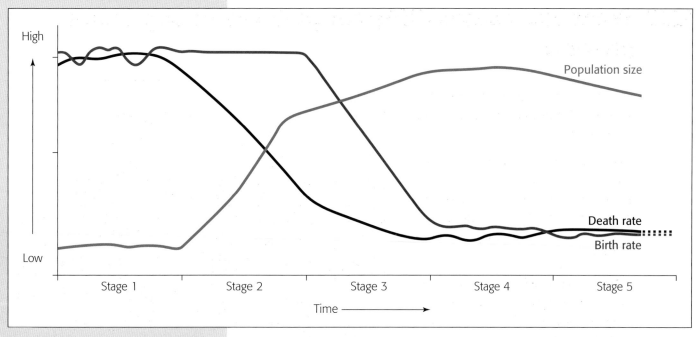

Figure 3 The demographic transition model

Population facts

90% of the world's births and 77% of the world's deaths took place in LEDCs in 1998.

99% of the world's natural increase takes place in LEDCs.

Geographers have produced a **model** to try to show these ideas. It shows the changes in population over the years. The technical term for this is a **demographic transition model**.

People live longer because we have managed to improve health and sanitation in many areas of the world. We understand the causes of many diseases and can cure them, or have managed to wipe them out (e.g. smallpox). We have also reduced the number of accidents that kill people. However, the AIDS epidemic is killing many younger people and drugs to treat the disease are expensive. Only wealthy developed countries can afford them. 63% of all AIDS patients live in sub-Saharan (Southern) Africa. The life expectancy in some of these countries is now getting shorter.

Over to you

1 On an outline copy of the demographic transition model (**Figure 3**):
  a shade in all the areas where the world population will grow
  b shade in all the areas where the world population will become less
  c mark on the point where the population grows fastest
  d mark on 2 places where it grows slowly.
2 The words in the boxes below describe a stage in the model.
  a Copy them in the right order under the appropriate section (stage) of the diagram.

Birth rate even lower	High birth rate	Birth rate getting lower	High birth rate	Birth rate quite low
Death rate stays low	High death rate	Death rate stays low	Death rate getting lower	Death rate low
Population begins to get less (decrease)	Population stays same (stable)	Population continues to increase	Population begins to grow quickly (increase)	Population stays same (stabilises)

  b Which stage of the model do you think the family in **Figure 1** fit into?
  c Which stage of the model do you think the family in **Figure 2** fit into?
  d What changes have taken place between 1904 and 2000?
3 Death rates are sometimes higher than normal. This may be the result of disease or accidents, and also famine, wars or natural disasters (earthquakes, drought, hurricanes and flooding). Use the internet to find out about recent deaths (e.g. in 1999, 2000 or 2001) caused by one of these.

POPULATION DYNAMICS

2.3 Population change in Malawi

Remember Abi, from page 109? Malawi, in East Africa, is Abi's home. Malawi is an LEDC. On page 107 we learnt that the total population of Malawi is likely to grow from 10 million in 2000 to 14 million by 2050. On pages 108–109 we saw how the number of women, who can have babies, has a big effect on the numbers of children born. The numbers of babies that they have also affects the population growth.

Population structure

Geographers need to look at the numbers of boys and girls, men and women, and the older people in a population. This is called the **population structure**. They study the structure of the population to predict for the future. The politicians and decision-makers need to plan for the needs of the people in 5, 10, 20 or 25 years' time.

To do this they draw up a population pyramid. **Figure 2** shows the population pyramid for Malawi. It shows all the people in Malawi in groups according to their age. Abi, who is 15, is counted in the 15–19 age band. This is called the 15–19 **cohort**.

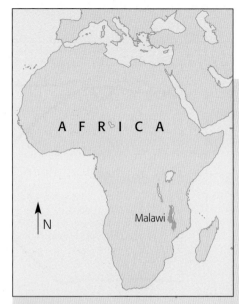

Figure 1 Malawi in East Africa

Figure 2 Malawi's population in 2000

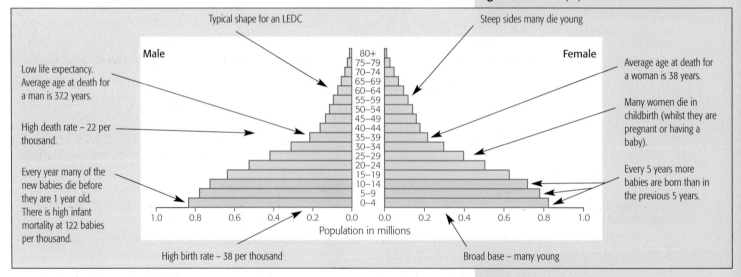

Malawi - the future

Abi filled in a quiz in a magazine about what things would be like in 2020. **Figure 4** shows Abi's responses to the quiz (in *green*). It also shows Moritz's response (in *red*). Moritz is 15 and lives in Germany, an MEDC. We'll look at Moritz's responses again on page 117.

Figure 3 Malawi's predicted population for 2050

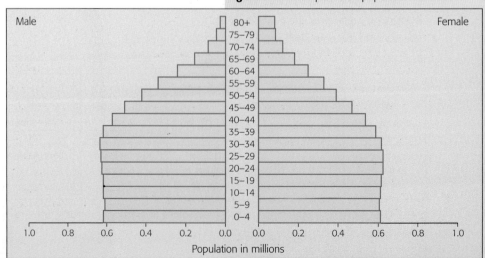

Study 2 Why is population changing?

WHAT WILL IT BE LIKE IN 2020?

How old will you be in 2020?	34 34		How far is your nearest clinic for advice?	Less than 1/2 hr. away ✔
				1/2 to 2 hours away ☐
Will you be	married ✔ with a partner ☐			2 to 5 hours away ✔
	widowed ☐ divorced ☐ single ✔ ?			1 day's journey away ☐
How many children will you have? 4 0	How many grandchildren? 1 0		Will you have your children vaccinated against childhood diseases so they won't die young?	Yes ✔ No, there are none available ✔
Social factors				No, it may not be good for them ☐
How many children do you think a woman should have?	<2 ✔ 2 ☐ 3 ☐ >3 ✔			No, I can't afford it ☐
Will you discuss family size before you marry?	Yes ✔ No ✔			They are more likely to die of AIDS ✔
			Career and economic factors	
Who will decide the family size?	You ☐ Your partner ☐ Both of you ✔ Your parents ☐		Do you think that a woman should have a career?	Yes, all the time ✔
	Your friends ☐ Your work colleagues ☐ The government ☐			Yes, but not when the child is young ☐
	Your church leaders ☐ It will just happen ✔			Yes, if child care can be arranged ☐
Health education factors				No, they should be at home to look after husband and young children and grandparents ✔
Have you been taught about family planning?	In school ✔ at clinic ☐ by doctor ☐			It is up to the husband to decide ☐
	by witch doctor ☐ by parents ☐ not at all ✔			
How easy is it for women to get contraceptives?	They can get them ✔		Which is more important to you?	Enough money to provide all you and your child want ✔
	They can get them free ☐			Enough money to live comfortably ☐
	They can buy them quite cheaply ✔			Children to look after you when you are old ✔
	They are far too expensive for most people ✔			Children to earn money for you now ✔
	They are not available anywhere nearby ☐			Children to work on the farm ☐
	I wouldn't know how to buy them ✔		Green Abi's responses	Red Moritz's responses
	My husband has the money not me ✔			

Figure 4 Is your future bright? How Abi and Moritz imagine theirs.

Over to you

1. On a blank quiz sheet, fill in your own answers to the questions.
   a. How do your answers compare with Abi's?
   b. How many children do you think Abi will really have? Can she really choose? Use her other answers to help you explain.
   c. How will Abi's family size affect the number of people in Malawi in 2050?

2. Look at the population pyramid for Malawi in 2000 (**Figure 2**).
   a. In 10 years' time which cohort will Abi be in?
   b. How many people do you think will be in the cohort with her – men and women? Explain your answer.

3. Now look at the predicted population pyramid for Malawi in 2050, **Figure 3**.
   a. Add annotations to describe the population pyramid.
   b. How is the birth rate different from in 2000?
   c. Approximately how many people now need jobs (aged 15–64)?
   d. How many people are now retired?

4. Group activity.
   In groups of four plan for 2000–2050, in Malawi.
   a. Choose a Minister for Education, a Minister for Health, a Minister for Employment and Pensions, and a Minister for Trade (to export produce, but to make sure there is enough food first).
   b. Each Minister should consider the following questions.
      - What is your department responsible for in order to provide for the population?
      - Will the changes mean that you need to provide more or less?
      - Will there be more or fewer babies? What facilities do babies need?
      - What will 2–4-year-olds need?
      - What should the government provide for the 5–14-year-olds?
      - What do the 15–64-year-olds need?
      - How and what will you provide for the 65+ age group?
   c. Your Government does not have enough to provide everything.
      i) Come to a group decision about your needs.
      ii) What do you think will be the most important to provide? Make a numbered list of your priorities. Explain your reasons.
      iii) In what other ways might you obtain money (funding)?

POPULATION DYNAMICS

2.4 How does migration affect Malawi's population?

The number of people who live in a country also changes as a result of **migration**. When people move away from one country to live in another, they are called **migrants**. **Figure 1** shows some migrants. Geographers divide them into different types, which are based on their reason for moving.

During the 1990s Malawi took in over 1 million migrants. Many of these have since either returned home, or settled in Malawi. In 1999 there were 1700 **refugees** in Malawi and 1300 **asylum seekers**.

■ More people

On 6 March 2000, 470 more refugees entered Malawi. Most come from neighbouring countries that are at war. Many men stay behind to fight or may already have been killed. The refugees can bring very few possessions with them and they have no job or home to come to. In Malawi at present the law says that refugees may not work. One refugee is a doctor and another is an airline pilot. The Malawians plan to change their law.

The United Nations try to keep a record of all the refugees but their data in Malawi does not include all migrants as many have settled and mixed with the local population. **Figure 2** shows the different age groups of the refugees at the end of 2000. Three out of every 1000 people in Malawi are refugees.

■ More to provide for

Malawi is small and economically poor. It does not have much agricultural land. It has taken in these migrants and settled them in a camp. There they rely on the forest to find food and to provide wood for homes, cooking, and heating. Malawi is losing too much forest land. (See also pages 120–121.)

Figure 1 What were the different reasons for these people becoming migrants?

Age	Male	Female
0–4	60	60
5–17	210	250
18–59	430	620
60+	40	40

Figure 2 Numbers of refugees in Malawi at the end of 2000, showing age and gender

Figure 3 People of all ages and backgrounds can become refugees

Study 2 Why is population changing?

Country of Origin	Number of families	Number of people
Somalia	55	205
Congo	279	697
Burundi	175	450
Rwanda	331	1023
Uganda	1	1
Eritrea	7	9
Sudan	6	9
Ethiopia	1	1

Figure 4 Details about refugees living in Dzaleka camp in May 2000

Figure 3 shows refugees like those arriving at Dzaleka camp. **Figure 4** tells you about where the refugees have come from. They are people of different ages. Many leave their homes and try to flee as a family but some of them die on the way.

Figure 5 is a newspaper report written about Dzaleka Refugee Camp.

A few more arrived...
Great-grandfather receives a warm welcome

31/5/00   Report from Dzaleka Refugee Camp

LAST MONTH one of Dzaleka's newest refugees was a 75-year-old great-grandfather all the way from Rwanda. He said 'I am very happy to be here, at peace at last.'

The transit shelter, now managed by Malawi Red Cross, has a full-time officer on duty to receive refugees as they come into the country. This May, 158 asylum seekers have passed through the shelter, hoping to find lodgings at the camp.

There are very few spaces available, even tents are hard to get.

Recently they have come from Democratic Republic of Congo but there are many in the camp from Rwanda, Burundi and Somalia.

It is illegal for the refugees to live outside the camp but many live in Lilongwe in order to find work or beg.

At last the Malawians have agreed to try and to provide education for these people.

However, this is difficult as not all Malawians have access to education themselves. In March, 17 youngsters joined a day secondary school just 6 km from the camp. The Malawi Red Cross provides uniforms, books and fees. A primary school has been constructed at the camp and local Malawian children also attend.

Activities are arranged in the camps to provide something for the people to do. Knitting groups have been set up for girls and women and one refugee has established discussion groups on behaviour, skills training etc.

Some young men set up activity associations in adult literacy, computer and accountancy skills.

The camp library is always well attended. JRS (Jesuit Refugee Service) work at the camp and try to arrange employment opportunities for the refugees in the local area so that they can be self-supporting.

Figure 5 Dzaleka Refugee Camp

Over to you

1 Read the newspaper article about Dzaleka refugee camp in **Figure 5**, and look at **Figure 3**.
   Imagine you are a refugee, and write a letter back to a friend in your home country. Tell your friend:
   a what it is like at the camp
   b what you miss most
   c what you are grateful for *and*
   d what you hope for in the future.
   Explain your ideas and give examples. You could sketch a picture of the camp.

2 Use an atlas to find the countries mentioned in **Figure 4**. Measure how far it is from each country to Malawi. You could find out about one of these countries, and why people are leaving (see also page 139).

3 Find the population pyramid for Malawi in 2000 on page 112. Draw another pyramid with the refugees from **Figure 2** marked on it.

4 Draw a cartoon for a Malawi newspaper. You will probably need to draw several pictures. Try to show the impact of all these refugees on Malawi.
   a What do the under-5s need?
   b What should the Malawians try to provide for the 5–17-year-olds? Do you think they can do this? Check in **Figure 5**.
   c Do you think that the 18–59-year-olds will be much help to the Malawians? Will they add more to the population by having more children?
   d What demands do those aged 60+ put on the population?
   e Can you suggest what you think the Malawians feel about refugees?
   Show your cartoon to a friend in the class and explain what you have drawn and why.

5 Do you have refugees who live in your area? What do local people think about them? Is it a fair opinion to have?

POPULATION DYNAMICS

2.5 Population change in Germany

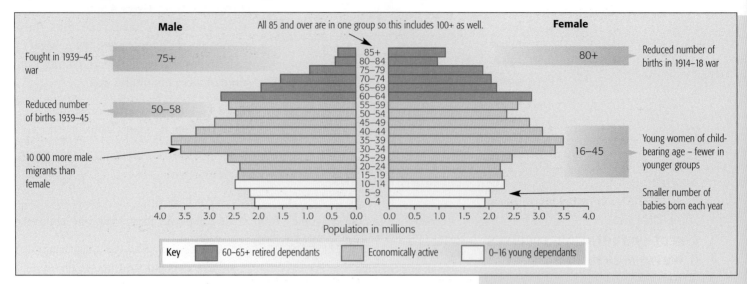

Figure 1 Population pyramid, Germany 2000

Germany is an MEDC and in this section we will consider why the population structure is changing in MEDCs. As we saw on pages 110–111, in MEDCs people like those in Victoria Brown's family are having fewer children. They are also living longer and more celebrate their 80th, or even their 100th birthday.

An example of a population pyramid for an MEDC is shown in **Figure 1**. It is for Germany in 2000. The annotations help to describe the population structure.

Figure 2 shows the **population processes** that are important in Germany. Notice the difference between the births and deaths. How does this compare with Malawi?

One of the reasons that the population is changing is that old people are living longer. **Figure 3** suggests some reasons why the people of Germany are now able to live to an older age.

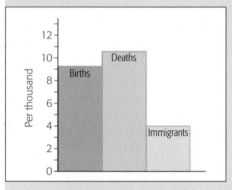

Figure 2 Changing population processes for Germany 2000

Figure 3 The ingredients for a longer life. Are all these things available to Malawians?

Study 2 Why is population changing?

■ Migration into Germany

In Germany many people move into and out of the country each year. Some are political asylum seekers from places like Kosovo and Bosnia. Others came immediately after the earthquake in Turkey in 1999.

As the population of Germany gets increasingly older and there are fewer people of working age, Germany will need more and more people to move into the country to work. They will not have enough people of working age (the **economically active population**) to fill all the jobs. Recent newspaper reports have commented on immigration in Europe (**Figure 4**).

Figure 5 Impacts that migrants have: which bring benefits and which could cause problems?

- May increase workforce
- Many have good skills
- May accept lower wages
- More need houses
- May need benefits
- May have **dependants**
- Pay taxes
- Money sent to home country
- Fewer to feed in home country
- Increase skilled workforce if return to home country
- May lead to racial tension
- May do unpleasant jobs
- May fill skills shortages
- May work long hours
- May start new companies and employ others
- Fewer young people in home country which then suffers **ageing population**
- Loss of young economically active in home country
- Loss of ambitious youngsters in home country

'Timebomb' alert as births tumble

Now a birth dearth in Europe. Fewer babies born in Europe than at any time since Second World War. European population increased by 266 000 in 1999. In India the population increases by 266 000 in about 6 days!

The Times 2000

The Smugglers' Routes

Illegal immigrants come into Germany from Iraq and Turkey.

The Times 3/11/2000

Immigrants are crucial to our continued economic wellbeing

Germany needs:
- 344 000 immigrants per year to keep the same population size.
- 487 000 to keep the same number of working people in the country.
- Over 3½ million to keep the same dependency ratio.

The Times 15/6/2000

Figure 4 Recent newspaper stories about population

Over to you

1 a Look at **Figure 1**, the population pyramid for Germany in 2000. Which of the following comments are true?

- There are more women than men.
- The biggest age cohort is 0–4.
- The pyramid gets smaller as it goes up.
- More people are over 60 than under 15.
- There are always more women than men in each age group.
- There are more older men than older women.

b For each comment that is untrue, write a true comment.

2 a What does **Figure 2** show about population change in Germany?

b Look at **Figure 4** on page 113 (the magazine quiz). The red answers were filled in by Moritz, aged 15, from Germany. Make a list of reasons why you think Moritz and his friends will have fewer children than Abi in Malawi.

c Compare your own answers to the quiz on page 113 with Moritz's. Explain the similarities and differences.

3 Figure 3 suggests some reasons why Moritz and his family already expect to live longer than Abi's family. For each picture write about 4 words to explain why they are less likely to die young.

4 Look at **Figure 4** and make a list of reasons why people might move to Germany in future. Do you think that these news reports are valid?

More information about migrants in Germany can be found at www.uni-bamberg.de/~ba6ef3/d

5 You have been commissioned to design the front page for a website. The site will allow people in Germany to find out in more detail all about the impacts that immigrants have. Look at **Figure 5** which shows some of the impacts. Use the table below and group the different topics into the appropriate boxes. Give the website users the sub-headings to click on. You can add pictures to make the site more interesting.

Impact on country	Benefits	Problems
Germany		
Home country they left		

POPULATION DYNAMICS

Study 3
The implications of population change

The world's population is growing, and the structure of population is changing. LEDCs such as Malawi have growing numbers of young people. MEDCs such as Germany have increasing numbers of older people – their population is ageing. What does this mean for the countries? And how are they going to cope with these changes?

Figure 1 How LEDCs have tried to encourage smaller families

3.1 LEDCs – coping with growing numbers of young people

In Malawi people see their children in a different way from parents in MEDCs. All parents value their children, but it is important for Malawians to have many children so they can help to support the family. As we saw on page 109, in most LEDCs very young children work to help the family. In MEDCs parents tend to have fewer children because they are expensive to look after.

Malawi has so many young people and, as they become parents, they too have many children. Some people think that it would be easier if there were not so many extra children born each year.

Abi's answers to the quiz on page 113 show that she has very little choice about her family. Her opinions are considered less important than her husband's. In MEDCs women can make more choices for themselves and are more independent.

Study 3 The implications of population change

They then choose to have smaller families. If Abi had greater choice and more independence she too would probably make different decisions. Some of the changes that might help Abi are shown in **Figure 1**. They have been tried in other LEDCs.

Throughout the world as countries have become more developed, people have changed their opinions and have begun to have smaller families. These changes are shown in the Circle of Development, **Figure 2**.

Meanwhile, the large number of children in Malawi causes difficulties for the Malawian Government. These children, just like you, ought to receive a good education. They also need enough food and free health care. Most of their families are farmers and live in poverty. The children want to be able to make the most of their lives. When they are older they will need jobs. They probably hope to go to the towns such as Lilongwe and Blantyre to work in factories. They want to earn enough to look after their own families in the future.

Figure 2 The Circle of Development

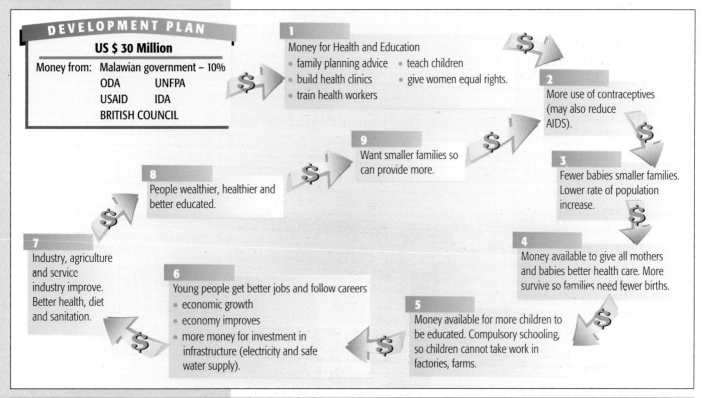

Over to you

1 **Figure 1** shows various ways that LEDCs have tried to encourage smaller families.
   a Make a list of the types of project which might help to reduce the numbers of children in LEDCs.
   b Make another list of the different ways that these projects can be paid for (funded).

2 **Figure 2** shows the Circle of Development. The stages are numbered 1–9. Draw a cartoon strip to show the circle of development in pictures so that the young people of Malawi could learn about it.

3 Make a list of all the things that the government should provide for the children in Malawi. Pair your list with the ideas below.
   - Foreign companies create new jobs by opening factories in towns.
   - Charities encourage local rural industries and self-help.
   - Free primary education for all children.
   - Parents encouraged to educate girls as well as boys.
   - Improve the land to grow food.

POPULATION DYNAMICS

3.2 Developments in Malawi

Aware of the need to limit the number of babies being born in Malawi, the Malawian government has tried to set up a number of projects. The government hopes to move the country further along the Circle of Development, which will encourage people to have fewer children.

The government is supported by the United Nations, USA, the UK and other organizations and charities. It has $30 million to improve conditions in Malawi. Each project hopes to reduce the number of babies born and lower the population growth rate. It also hopes to provide an adequate **standard of living** for the young.

Figure 1 shows other projects funded by the World Bank. The World Bank gives money from MEDCs to support projects in LEDCs.

Figure 1 Projects in Malawi funded by the World Bank

Project funded by the World Bank	Year begun	Money available US$millions
Secondary Education Project	1998	48.2
Social Action Fund 1 & 2 (social protection)	1998 and 1996	66 and 56
Population and Family Planning Project	1998	5
Primary Education Project	1996	22.5
National Water Development Project (water supply and sanitation)	1995	79.2
Agricultural Services	1993	45.8 (now closed)
Power Project (electricity)	1992	55 (now closed)
Population, Health and Nutrition Sector Credit Project	1991	55.5

Figure 2 What's most important? Village people in Ngulukira meet to decide

WANTED IN NGULUKIRA
Projects by and for local people.
1 New school – 5-15 yr olds. BUILT
2 Clinic + health advisor. SET UP
3 Advice centre for farming and fishing. MEETS MONTHLY
4 Boreholes (wells) for safe water – avoid disease. DRILLED + WORKING

PLANS
5 Better policing – by locals
6 Further development to be led by local people – appoint co-ordinator
7 Investment in smallholder farms
8 Better opportunities for women. They should make decisions and control their lives.

Study 3 The implications of population change

Figure 2 shows the people of Ngulukira, a village in Malawi. They are at a village meeting to discuss the projects that they want in order to develop their village. They have been given money by a charity from the USA.

Figure 3 is a report from a TV programme about Malawi. It tells of projects to improve the environment. These changes will then help the poor farmers and their young families.

PLANT A TREE IN MALAWI

All the extra young people in Malawi need food. So each year the farmers clear more forest to grow crops. The women collect more firewood to cook food in their homes. But as the trees are cut down, the soil is eroded and the land soon becomes infertile. Then fewer crops grow and the land becomes waste land.

CPAR (Canadian Physicians Aid Relief) is a charity based in Canada. They have launched a campaign to 'plant a tree' to replace the forest and reuse the waste land. These trees will provide a variety of food crops. Crops can be grown in their shade and the soil will not be eroded.

CPAR has 75 workers in Malawi. 15,000 trees were planted in one year. Most trees are given to schoolchildren to plant around their schools and villages. CPAR also raises money to provide health and education facilities for the young people of Malawi.

Figure 3 How planting trees in Malawi can help with producing food and improving the environment

Over to you

1. Look at **Figure 1**, showing projects in Malawi funded by the World Bank.

    a Complete the table below:

Project type	Total amount spent
Health and Family Planning	
Education	
Cleaner water and sewerage	
Better agriculture (Food)	
Electricity to improve living conditions	
Social projects (e.g. women's rights)	

    b Draw a bar graph to show the amounts spent on each type of project.

    c What do you think that the graph shows about development in Malawi?

    d Look back at the Circle of Development on page 119. On your graph add on each bar the numbers of the stages that the World Bank money will help. Are all the sections provided for? Would you give funds for anything else? Why?

2. **Figure 2** shows the meeting in the village of Ngulukira in Malawi. The villagers have been given funds by USAID, a charity from the USA. The villagers have met to discuss progress so far.

    Write a report to the USAID charity co-ordinator.

    a Tell her what has been achieved so far in Ngulukira. Explain why it is helping the village.

    b Rank the 4 new plans in your order of priority.

    c Justify your order and explain how each project will help the local villagers.

    d Think of 2 more plans that you think the villagers might like to consider to improve development in future. Think about the Circle of Development. (Hints: move jobs from agriculture to industry? better services for local people?)

3. Read **Figure 3** 'Plant a tree in Malawi'. CPAR and other organizations (e.g. UNHCR) have provided trees for Malawi. Explain how they are helping to provide a better life for the young of Malawi

4. Use the ideas and answers on pages 118–121 to write an answer to this exam-type question:

    Using examples from places you have studied, explain how LEDCs cope with the demands of an increasing population.

3.3 MEDCs – coping with an ageing population

> 'One of the major achievements of the 20th Century has been an increase in life expectancy in almost every country in the world.'
> Helpage Charity.

Between 1995 and 1996 there were an extra 12 million people aged 60+ in the world. All of the countries in the world will in future need to make decisions about how to cope with older people. Many MEDCs are already aware of the extra number of elderly people they have. This also means of course, because of lower birth rates, that they have far fewer young people.

By 2050 the population pyramid for Germany will look like **Figure 1**. You can compare it with **Figure 1** on page 116.

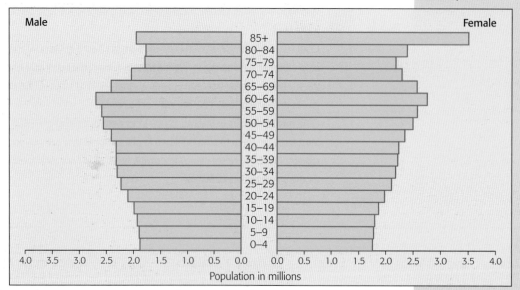

Figure 1 Population pyramid for Germany, 2050

Providing for elderly people's needs

Elderly people are often very independent and they live very active lives. **Figure 2** shows some statistics about the elderly population in Germany.

Figure 2 How it adds up for the elderly in Germany

14% of elderly live **with their own grown up children**.

41% of elderly live **alone**.

Dependency ratio in Germany in 2000 = **2.1**
(About twice as many dependants as economically active.)

Average age of **retirement** is **58.4 for women** and **60.5 for men**. Pensions paid from 61 years old.

German **pension** almost all paid by government. Very generous – **88%** of employed pay.

After age 60 only **2.9%** are in **paid work**. By 2030 this will probably be **5.5%**.

Germans spend **2.7** times as much on **health care** for those aged 65+ as for all those aged 0 to 64.

3.5% of Gross Domestic Product (**GDP**) is spent on health care for those aged 65+.

Study 3 The implications of population change

Figure 3 Looking at this evidence, what could you say about elderly people's needs?

Governments and society need to be certain that elderly people have a good quality of life. Old people do not usually have a job and so they are not able to earn any money to live on. Some of them may have savings but they do not know how long they will live. They may need to make their savings last a very long time.

In most MEDCs the government gives older people a pension when they retire. The government has to raise this money from taxes from those who are earning. By the time that you retire there will be very large numbers of elderly people. The government is already concerned about your pensions. It will not be able to afford to pay pensions to all of you. You will have to have a private pension, one that you pay for yourself. You will need to start saving as soon as you get a job.

Some people think that older people place very big demands on the economically active people who pay the taxes. **Figure 3** shows some of the demands that they make.

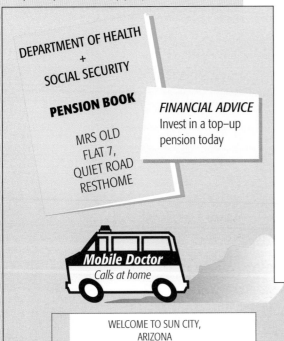

Over to you

1 a Annotate the population pyramid for Germany 2050 to show the main characteristics of the population.
  b Add comments in a different colour to show the differences from 2000 (**Figure 1**, page 116).
2 Calculate the **dependency ratio** for Germany in 2050.

$$\frac{\text{Dependants}}{\text{Economically active}} = \text{Dependency Ratio}$$

What does it mean?

3 Make a list of all the things that you think a retired person might want to spend money on.

  a Is the list the same as a list that you might make?
  b What is different? Why?
  c Is the list the same as your parents might make?
  d What is different? Why?

4 Look at the demands suggested in **Figure 3**.
  Write a letter to your MP. Say what you think the government should pay for and what you think they should expect the elderly to do to help themselves. (Remember that you will be elderly one day!)

POPULATION DYNAMICS

3.4 Valuing our elderly population

Throughout the world, many people can expect to live for 20 or 30 years after they retire. They will probably be very fit and active. They may not move as quickly as when they were younger but they will have plenty of experience to bring to any situation.

They will want to keep busy and be involved in activities. **Figure 1** shows some of the things that one elderly person, Peter Gray does.

As we can see, people like Peter often work just as hard after they have retired as they did before. But now they aren't paid for their work. They are **volunteers**. They may work for family, for friends or for the community.

Figure 1 Peter Gray's diary

WEEK November 15–21		WEEK November 15–21	
15 MON a.m.		**FRI** a.m. Work for Charity 8.45–12.	**19**
p.m. Collect Freya from school/clean up home.		p.m.	
16 TUES a.m. Drive minibus for volunteer hospital transport/visit Graham in hospital.		**SAT** a.m. Shop for Mrs. Cook. Collect pension and fuel grant forms. Book May holiday.	**20**
p.m. Flu vaccination		p.m.	
17 WED a.m. Course for University of Third Age 9.00–12.00		**SUN** a.m. To Katy for lunch.	**21**
p.m. Parish Council Meeting 8.00p.m.		p.m.	
18 THURS a.m. Deliver Meals on Wheels 11.30–2.00		**OTHER** Write to MP about safer cycle paths. Arrange new bus pass.	
p.m.			

Figure 2 shows an OXFAM distribution centre with one of its workers sorting clothes. Some workers are retired. This is similar to the work that Peter Gray does on Fridays.

Figure 4 is an article from a local newspaper. It shows the value of older people to a school in Essex. Older people are often valued for the experience they gained during their lives. They can pass on this knowledge to younger people. It is far more fun to listen to someone's story rather than to read about it in a book!

Figure 2 Retired people may work as volunteers for charities

Study 3 The implications of population change

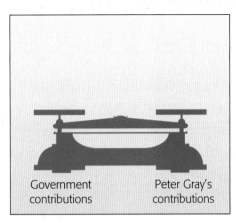

Figure 3 Weighing up the contributions

Figure 4 Older people sometimes go back to school!

Lessons from those who were there

OLDER people delighted in sharing their wartime stories with youngsters at a specially-arranged tea party.

Year five pupils at the Leigh Beck school in Point Road, Canvey, invited 40 guests as part of their Britain in the 1930's project. Teacher assistant Susan De'ath said: 'It went very well indeed and the children were enthralled with the experiences they heard. They listened with great interest.'

Children served tea and biscuits before interviewing the veterans where they asked questions about the war.

Mrs De'ath said: 'I think our visitors liked sharing their experiences with the children. Many of them brought some medals and mementoes with them too, which thrilled the kids.'

Over to you

1 Look at the diary of Peter Gray, **Figure 1**. List all the unpaid jobs that he is doing this week (Katy is his daughter, and Freya is his granddaughter).

2 Pair up each of Peter Gray's activities with one of the ideas below.

He is helping the economically active:
   i) to do their own jobs
   ii) to reduce the cost of care for the sick
   iii) to reduce the cost of childcare
   iv) by doing the jobs that don't pay well.

He is helping in the community:
   i) by representing the people
   ii) by campaigning for better conditions
   iii) by looking after the poor.

3 For some of his activities he will probably get government payments. Pair up the activity with the support below.

He needs support from the government for:
   i) income/money
   ii) heating payments
   iii) education fees
   iv) transport costs
   v) healthcare
   vi) reduced charges at events.

4 Copy the diagram of scales in **Figure 3** for Peter Gray's and the government's contributions to elderly people.

   a Use the ideas in question 2 and draw a box on the weighing scales to show each contribution. Draw the size of the box to represent what you think is the value of each contribution – bigger contributions, bigger boxes. The contributions that Peter Gray makes go on the right of the scales.

   b Use the ideas in question 3 above and add boxes to show the contributions that the government makes. These boxes go on the left of the scales. The bigger the government contribution, the bigger the box.

   c When you have finished, draw an arrow to show the way that the scales will tip. Who makes the bigger contribution?

   d Does Mr Gray depend on others or do others depend on him?

5 Read the newspaper article, **Figure 4**. How did the elderly help at Leigh Beck school? Think about all the ways they helped – helping teachers, helping children, providing visual aids, etc.

6 Interview a retired person that you know. What do they do in their spare time? What do they need help with?

7 Find out what organizations like Age Concern and Helpage do to help the elderly.

8 In what ways can you help in your local community? Does your school have links with the elderly in your community? Are there other youth organizations that help?

POPULATION AND RESOURCES

Study 1
What are resources?

In this study you will find out what resources are. You will think about how we can classify them into different types, and how they are used.

1.1 What are the differences between different types of resources?

In this section we want to find out about resources and how they are used. The photographs in **Figure 1** show different people using a wide variety of resources in different ways.

■ So, what are resources?

Resources are anything that people can use.
They are all the things we need to live and work. A variety of possible resources is shown in **Figure 2**.

> Different tastes (snails are a French food delicacy), different lifestyles, and different skills use different resources. Resources are different things to different people.

Some resources are being used up and cannot be replaced (**non-renewable**). Others can be grown and so replace what has been used already, or used again and again (**renewable**). We can group resources according to whether they will run out or can be replaced or be used again and again. This is called **classification**. **Figure 3** shows a variety of resources and their classification.

> Resources can be classified, or grouped, as either renewable or non-renewable.

Figure 1 Different resources are used in different ways

Study 1 What are resources?

Figure 2 What are resources?

Renewable	Non-renewable
Solar power	Oil
Lambs	Copper ore
Oranges	Iron ore
Hazelnuts	Limestone
Wheat for flour	Aluminium from bauxite
Milk	
Flowers	
Grass	
Wind power	
Water	

Figure 3 Renewable and non-renewable resources

Over to you

1 Look at **Figure 2**. It shows a variety of resources.
   a Which of the items would be a useful resource for you at home or at school today? Explain why.
   b Which would be useful to you at other times? (e.g. camping, at the weekend, at friends' houses and so on.)
   c Which would you never be able to use on your own? Why?

2 On a copy of **Figure 3**, add some further information. Use the items already in the table as a guide to where to put others.
   a Add the following resources to the table: coal, maize, apples, chicken, wave power, granite rock, lettuces, silica sand, vegetable oil.
   b Put the following phrases under the correct columns: animal products, fuels, finite (will come to an end), can be re-used, cannot be replaced, plant products, can be replanted or re-grown, minerals, will run out, recycles naturally in the environment, metal ores.
   c Write out your own definition of 'renewable' and 'non-renewable'. Use the phrases in the chart to help, and give examples.

   d Do all resources fit neatly into one group or the other? Where would you put the sun's energy, soil and uranium (a small amount can last for hundreds of years)? Can you think of any other problems with classifying resources in this way?

3 Look at the photographs in **Figure 1**.
   a What resources are used in each photo? Make a list for each photo.
   b Are some resources used in all photos?
   c Would the resources in photo **a** be useful for the people in photo **b**?
   d Would the resources in photo **b** be useful for the people in photo **a**?
   e What do these photos suggest about the usefulness of resources?

POPULATION AND RESOURCES

1.2 Using resources

As we saw on pages 126–127, different people use different resources in different ways.

The people who live in the rainforest in photo **b** on page 126 make almost all that they need from resources that they find in their own part of the forest. Their weapons are made from wood, used for arrows and blowpipes, and plant juices provide the poison. They do not use gas or electricity. They also use wood for building houses and boats. They use plants for house roofing and food.

The Indian farmer in photo **c** on page 126 uses local products. He uses a wooden or metal plough. He cannot use a tractor or electricity.

In the photograph of the Frankfurt Stock Exchange, photo **a** on page 126, they are using computers, desks, and shelves. These items are made from plastics and metals but wood is also used. They obtain the resources from all over the world.

Some resources, such as wood, are useful for many purposes and many people. Some, such as plastics, may become more useful in the future as we develop the technology to use them for more things. Other resources, such as straw for roofing, are now less valuable than they used to be. We have replaced them with something else.

Figure 1 shows a range of energy resources and their various uses.

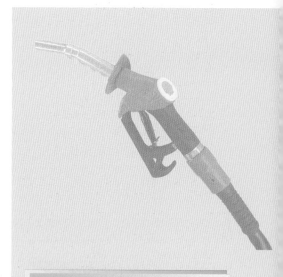

The usefulness of resources is constantly changing. Scientific and technological advances have led to a change in resource use.

Energy source	Uses	World supply for	Problems	Advantages	Renewal time
Coal	• fuel for heat • use in chemical industry • medicine	220 years	• produces smoke and soot • gives off CO_2 • heat efficiency only 30%	• used by many power stations all over world • known technology • not too expensive	millions of years
Wood	• fuel for heat • paper • ground cover (bark chippings) • building material • fibres from cellulose	could be forever but will need to be looked after	• use leads to deforestation • reserves may be lost • woodland may be needed for other uses	• inexpensive • widespread • very simple technology to burn	5–10 years for some, whole trees up to 100 years
Plant material	• fuel for heat • decompose to a gas for heat and light • food • animal fodder • fertilizer • animal bedding • traditional clothing (e.g. grass skirts)	probably forever	• bulky to transport, used for other things • lacks prestige as a fuel • efficiency varies • large quantity needed • land needed for other crops	• very cheap • available almost everywhere • uses waste • can be made into gas • less pollution than coal or oil • fairly simple technology	up to 1 year (may only be 4 months)
Oil	• can be divided into many types – for heat or lubricants • used to make plastics and synthetic fibres (nylon)	40 years – or will more be found? or something different?	• pollutes when burnt (CO_2, NO_x, SO_2) • heat efficiency 40% • environmental disaster if spilt at sea	• used all over the world • complicated technology developed already • MEDC lifestyle based on it	millions of years to compress small organisms deep inside the earth

Figure 1 Weighing up energy resources: how are they used and what are their advantages and disadvantages?

Study 1 What are resources?

When a resource becomes scarce, people may need to find a substitute. This is called **SUBSTITUTION**.

People are becoming increasingly aware of the impact that the exploitation of resources has on the environment, for example causing pollution and global warming. Resource use needs to be managed.

■ What can we do if resources run out?

Figure 1 also shows how long the world supply of the resource mentioned will last. Within forty years we will have run out of oil unless new supplies are found. People are already working to find alternatives for oil.

■ Resource exploitation

When resources are used or exploited they may cause problems for the environment. Other resources may need careful management if we are to use them well.

Figure 2 shows the problems caused by burning fossil fuels. You can find out more about these problems on pages 186–197.

Figure 2 Global warming and acid rain: two of the problems caused by burning fossil fuels

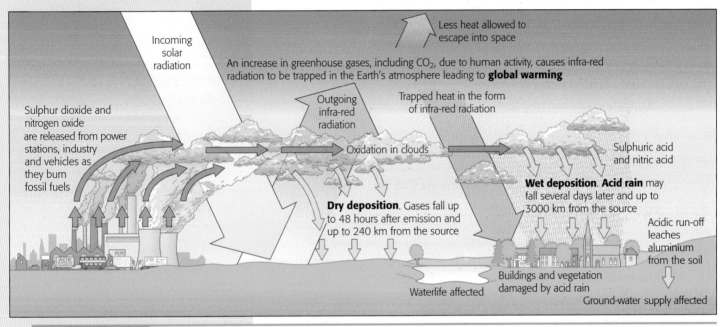

Over to you

1 Some resources are shown in **Figure 1**. They can all be burnt to provide heat and a little light but they are very different.

   a Which could be used in 2010 to produce:

   i) food   ii) clothes   iii) heat   iv) fuel for cars?

   b Which are likely to be used in 2050 for:

   i) food   ii) clothes   iii) heat   iv) fuel for cars?

   c Which are likely to be used in 2250 for:

   i) food   ii) clothes   iii) heat   iv) fuel for cars?

   d Which resource seems most useful for the future? Why?

2 Imagine that you live in a world where oil has run out. There will be no oil, diesel or petrol. Describe your day. Make it as similar to today as possible. What would have to be different? What do you think will be the substitutes for those things that today need oil and petrol?

3 You are going to take part in a debate. The motion is 'The world must stop burning oil. It is too precious to send up in smoke.'

You can be 'for' or 'against' the motion, but you need to prepare some notes for the debate.

Use these ideas to get you started.

- What is it useful for?
- Is it really precious?
- What would life be like without it? (Have we a substitute?)
- Can we manage without it today?
- How long will it last? (Is it **renewable** and **sustainable**?)
- What about the next generation?
- Should we save it for special uses? (**Conserve** it?)
- What other problems does it cause? (**Environmental impact**!)
- Can we reduce these problems enough?

   a In groups, or as a class, debate the motion.

   b You could debate the same motion but consider 'wood' or 'coal' instead of 'oil'.

POPULATION AND RESOURCES

Study 2
How are energy resources being used?

In the previous study we looked at the difference between renewable and non-renewable resources. This study looks at the use of energy resources in two different areas of the world, France and India. We will think about whether their energy supplies are **sustainable**, and look at their impact on the environment.

2.1 France, an MEDC

France (see **Figure 1**) is a **More Economically Developed Country** in Europe. It needs energy for farms, factories, shops, offices, for electrical equipment in homes, and for leisure activities. **Figure 2** shows the amount of energy consumed by different users in France.

Where does France's energy come from?

France has some coal in the north-east and central areas of the country. It has natural gas deposits in the south-west. But it does not have enough fuel to supply the large thermal power stations which could produce the energy it needs. What it does have, however, are high mountains suitable for hydro-electric power generation and also supplies of uranium which are used to produce nuclear power. **Figure 3** shows the types of power generation in France.

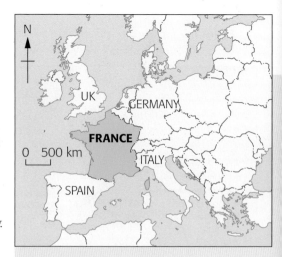

Figure 1 France

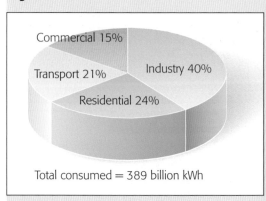

Total consumed = 389 billion kWh

Figure 2 Who uses energy in France?

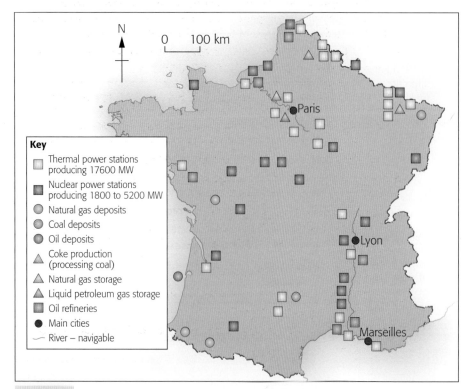

Type of energy	%
Thermal	11%
Hydro-electric	13%
Nuclear	76%

Figure 3 Power generation in France

Figure 4 Where thermal and nuclear power is generated in France

Study 2 How are energy resources being used

Figure 5 A nuclear power station at Nogent-sur-Seine, France

Thermal power stations, using fossil fuels, provide only 11% of all France's energy. **Figure 6** shows how a thermal power station works.

In 2001, 76% of France's energy came from nuclear power stations. France uses its own uranium in these power stations and also imports uranium. Nuclear power stations look and work very like thermal power stations but they use uranium (a nuclear fuel) rather than fossil fuels to heat the water to produce steam. **Figure 5** shows a nuclear power station at Nogent-sur-Seine, near Paris. Some people include them as one type of thermal power station. Both nuclear and thermal power stations are shown in **Figure 4**.

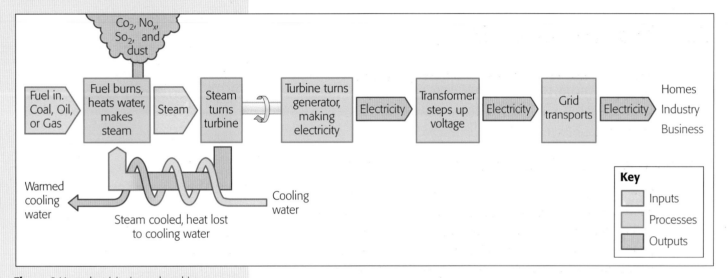

Figure 6 How electricity is produced in a thermal power station

Over to you

1. Look at **Figure 4**. Describe the distribution of thermal power stations in France. Think about the types of places they are found: coasts, rivers, near energy supply (coal, oil, gas), near big cities where many people live.

2. Look at **Figure 5**. It shows a nuclear power station. Describe what it looks like by copying the table below and circling the answer.

Site	Area	Height	Built of	Appearance
Flat/sloping	Large/small	Tall/low	Metal/brick	Eyesore/blends in

   You can find more on the website **www.edf.fr/htm**

3. Draw a sketch of the picture in **Figure 5** and annotate it to show the visual and other environmental impacts.

4. Look at **Figure 6**. It shows the inputs, processes, and outputs of energy production for a thermal power station. Draw three boxes and label them 'inputs', 'processes' and 'outputs'. Fill in each box with a list taken from the diagram.

5. Look again at the 'outputs' box.
   a. Tick the products that you think are useful.
   b. Cross the products that you think are harmful.
   c. Do you think that thermal power is **environmentally friendly**? Justify your answer.

POPULATION AND RESOURCES

2.2 India, an LEDC

In these two pages we will look at power development in India. India (see **Figure 1**) is a **Less Economically Developed Country** in Southern Asia. Only in the past 50 years, since gaining independence from Britain, have people begun to use larger amounts of energy. In the cities many people use energy in exactly the same way as Europeans. The recent development of industry has also led to more energy use. Energy use is now increasing at 8% per year.

Where does India's energy come from?

India has supplies of coal, uranium, gas, and oil. It also has mountains and large rivers which can be used for hydro-electric power. **Figure 2** shows the energy sources and power production in India.

In India the pattern of energy use is very different from in France. The amount of energy consumed by different users in India is shown in **Figure 3**. The two sections that are increasing the fastest are residential (household) and agriculture. In the rural areas there are many poorer people and they often rely on the sun for light and wood for fires for cooking. Instead of using machinery, much work is done by hand or by animals (see **Figure 1c**, page 126). Recently people have tried to use simpler technology to bring electricity to rural areas of India.

In this section we want to look at one type of energy resource in India. We will look at biogasifiers. **Figure 4** shows some areas that have biogas projects.

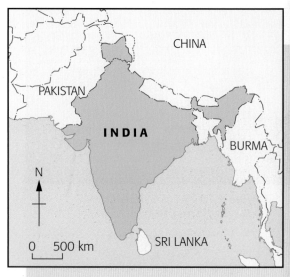

Figure 1 India

Type of energy	% share
Thermal	72%
Hydro	24%
Nuclear	3%
Wind	1%

Figure 2 Power generation in India

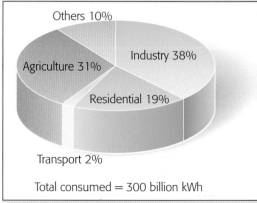

Total consumed = 300 billion kWh

Figure 3 Who uses energy in India?

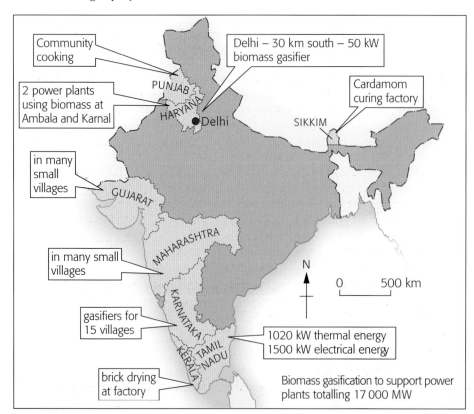

Figure 4 Biogas projects in India

Study 2 How are energy resources being used?

Figure 5 shows a biogas plant, whilst Figure 6 shows how biogas is made.

Figure 5 A biogas plant

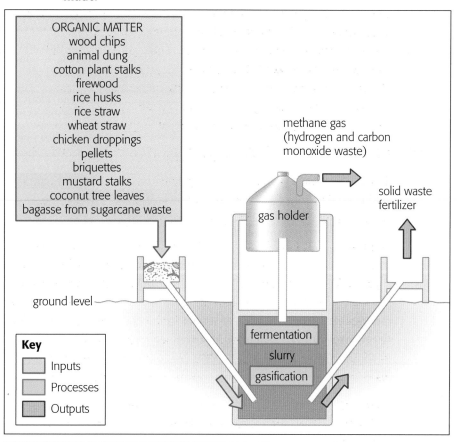

Figure 6 Biogas production

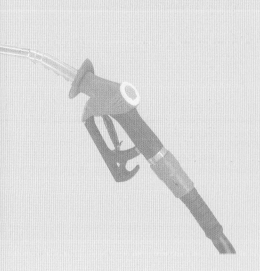

Over to you

1. Use the data in **Figure 2** on page 130, and **Figure 3** on page 132. Draw a bar graph to compare the amount of energy consumed by different users in France and India. Suggest reasons for these differences.

2. Look at **Figure 4**. Describe the distribution of biogas projects in India. Think about the types of places they are found. Are they near coasts and rivers, near coalfields, near farming areas, far away from big cities where many people live? You may need to use an atlas to help you.

3. Look at **Figure 5**. It shows a biogas plant. Describe what it looks like by copying the table below and circling the answer.

Site	Area	Height	Built of	Appearance
Flat/sloping	Large/small	Tall/low	Metal/brick	Eyesore/blends in

4. Draw a sketch of the picture in **Figure 5** and annotate it to show the visual and other environmental impacts of a biogas plant.

5. Look at **Figure 6**. It shows the inputs, processes and outputs of energy production for a biogas plant. Draw three boxes and label them 'inputs', 'processes' and 'outputs'. Fill in each box with a list taken from the diagram.

6. Look again at the 'outputs' box.
   a. Tick the products that you think are useful.
   b. Cross the products that you think are harmful.
   c. Do you think that biogas is environmentally friendly? Justify your answer.

POPULATION AND RESOURCES

2.3 How sustainable are these energy supplies?

In planning for the future, geographers need to know what impact the use of resources is having. They need to know the amount that remains for the future (the resource stock). They also need to know what effect the use is having on the environment (the air, water, soils, and plants). These are the questions they ask:

Will the resource still be available in 50 years' time?

Will the environment still be the same in 50 years' time?

In other words, is the resource use **sustainable?**

Sustainable is an 'in' word. It describes the way we can use resources. Some resources can be managed to make them sustainable.

People have begun to ask what will happen when resources run out. Some resources don't need to run out but if we don't use them well they will be exhausted and no longer available. If we take care, and replant trees and look after soils, they will last forever. Then our use of them will be sustainable.

Sustainable use really means that the resources we have today will be available for the next generation. What you and your parents use will still be available for your children and grandchildren.

■ Thermal power stations

Figure 1 is an e-mail conversation between a GCSE student and a French power station representative. They are discussing the environmental impact of the thermal power stations.

Figure 1 A GCSE student and a French power station representative discuss thermal power stations

---

Coal, oil and gas are running out — shouldn't you stop using them to make electricity?

>> We use them to provide for peak demand, to top up supply. Anyway, we've got the coal, we've got the buildings, we might as well use them. There's a lot of money tied up in all this.

But you're wasting a lot of the energy as well. It's only 30-40% efficient.

>> We're using combined heat and power now so that we use the heat from the steam instead of wasting it. We're running at 55% efficiency now in some power stations.

Getting the coal from the ground and the oil to the power stations causes so much environmental damage. The quarries and mines look awful and we keep on hearing of oil spills in the sea.

>> We have laws now to make us restore the land after we have finished quarrying and mines are making fantastic tourist attractions. The environment returns to normal eventually after an oil spill and we have plenty of ways of trying to clean it up.

But you are also polluting the air and causing global warming and acid rain. The sea level is predicted to rise 25—40 cm by 2040. Most of the CO_2 comes from power stations.

>> This has been a problem but we are using low sulphur content fuels, adding lime to old furnaces and installing flue gas desulphurization and denitrification systems to E.U. standards. We are even producing gypsum for the plaster industry from the waste now. 90% of the SO_2 is removed and 99% of the fly ash too. It's expensive and we've spent several billion French francs on this. We are doing our best!

I still think that the cost to the environment is too great. I hope that you will have better alternatives soon and that you can use the remaining coal for chemical and other purposes.

>> It will take time.

Study 2 How are energy resources being used?

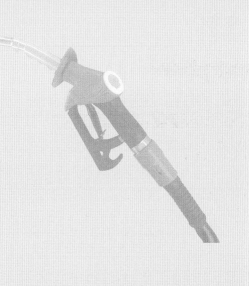

Figure 2 Environmental impact analysis

How sustainable is it?											
Will it run out?	−5	−4	−3	−2	−1	0	+1	+2	+3	+4	+5
Fuel source											
Non-renewable or renewable?	−5	−4	−3	−2	−1	0	+1	+2	+3	+4	+5
Pollution											
Does it pollute?											
■ Atmosphere – CO_2	−5	−4	−3	−2	−1	0	+1	+2	+3	+4	+5
SO_2	−5	−4	−3	−2	−1	0	+1	+2	+3	+4	+5
NO_x	−5	−4	−3	−2	−1	0	+1	+2	+3	+4	+5
■ Noise	−5	−4	−3	−2	−1	0	+1	+2	+3	+4	+5
■ Rain – acid?	−5	−4	−3	−2	−1	0	+1	+2	+3	+4	+5
■ Water – heated?	−5	−4	−3	−2	−1	0	+1	+2	+3	+4	+5
Is it radioactive?	−5	−4	−3	−2	−1	0	+1	+2	+3	+4	+5
Does it modify the landscape (visual pollution)?											
■ On extraction	−5	−4	−3	−2	−1	0	+1	+2	+3	+4	+5
■ When processing	−5	−4	−3	−2	−1	0	+1	+2	+3	+4	+5
■ When used	−5	−4	−3	−2	−1	0	+1	+2	+3	+4	+5
■ With waste deposits	−5	−4	−3	−2	−1	0	+1	+2	+3	+4	+5
Waste											
Is there a useful by-product? (name it........................)	−5	−4	−3	−2	−1	0	+1	+2	+3	+4	+5
Efficiency											
Is it efficient at producing energy from the resource?	−5	−4	−3	−2	−1	0	+1	+2	+3	+4	+5
Is it controllable, making energy only when needed?	−5	−4	−3	−2	−1	0	+1	+2	+3	+4	+5
Cost											
Is it costly:											
■ to produce?	−5	−4	−3	−2	−1	0	+1	+2	+3	+4	+5
■ to clean up?	−5	−4	−3	−2	−1	0	+1	+2	+3	+4	+5

Over to you

Figure 2 is an environmental impact assessment form. It can be used to assess the impact and sustainability of the thermal power scheme in France and the biogas scheme in India.
We will look at the thermal scheme here.

1 On a copy of **Figure 2**, complete the assessment form. Use the comments in **Figure 1**, and your work from pages 130–131.
   ■ For each category you need to award a mark (score) between −5 and +5.
   ■ If the resource use has **no effect** give it 0.
   ■ If the resource use has a **harmful effect**, give it a score between −1 and −5. If it is very harmful it will score −5 but if it is only a slight problem, it will score −1.

■ If the resource has a **good effect** it will have a score between +1 and +5. If it is very good it will score +5 but if it is only just good it will score +1, of course.

2 Calculate the total score for the thermal scheme. Do you think it is sustainable? Explain why?

3 France has been making many changes to try to protect the environment. Look again at **Figure 6** on page 131. Look also at the replies from the power station representative in **Figure 1**
   a What are these changes? List them under these headings: laws, efficiency, pollution, alternatives.
   b What do you think should happen in the future in France? Why?

135

POPULATION AND RESOURCES

2.4 Are biogas plants better than thermal power stations?

You have looked at the thermal power station in France, assessed its environmental impact and sustainability. Now we can look at the biogas plant and see whether in India they do any better! Are the biogas plants sustainable?

Figure 1 shows two UK gap-year students who are backpacking in India. They are talking to an Indian businessman. Like the GCSE student and French power station representative on page 134 they too are discussing environmental impact. This time it's about biogasifiers.

India is trying to make its energy supply more sustainable in other ways. It is reducing air pollution from its power stations, and it is trying hard to cut down the amount of energy lost when electricity is transmitted by electricity pylons. (Local biogas schemes do not transmit energy over long distances.) The Indian government and the World Bank have also begun to replant trees where forests have been cut down for fuel, and have introduced better fuelwood stoves that could save up to 70% of the wood used to cook a meal.

Figure 1 Biogas: getting the facts

I hear that you are still burning wood to produce your heat and also to make electricity. It's very wasteful. 90% of heat is lost when you burn wood on an open fire.

That's true but it's not quite what you imagine, I think. We are converting our biomass to gas in a gasifier. All the energy is converted to gas and then we can use 75% of the heat. We only lose 25%. We can also store the gas until we want it to make heat or to produce electricity and we've cut out all the smoke and the ash that our open fires used to create. It used to make our eyes sting and the children coughed a lot.

These gasifiers are very small though. Is it worth it? A large power station would generate hundreds of megawatts.

We do use the gas in some power stations to generate electricity for larger places but 'small is beautiful' for us. Many of us live in villages quite a long way from major roads and main towns and cities. Small means that it is affordable to us and local. We can control it and we are not dependent on big companies. Our village co-operative owns our biogasifier. The women and children are happier now because they do not have to collect firewood and that saves them so much time. The trees are looking stronger and healthier too!

So how do you get the fuel for the gasifier?

We can use all sorts of wastes that can be found in the village. Dung, straw, all sorts. We can find our fuel resources all over the place and each time we harvest a crop or feed the animals we get more.

Study 2 How are energy resources being used?

Over to you

On pages 134–135 we used an environmental impact assessment form to assess the impact and sustainability of the thermal power scheme.

We will now look at the biogas project.

1. On another copy of the environmental impact assessment form, complete the assessment for the biogas project. Use the comments in **Figure 1** and your work from pages 132–133.
   - Remember that for each category you need to award a mark (score) between −5 and +5.
   - If the resource use has **no effect** give it 0.
   - If the resource use has a **harmful effect**, give it a score between −1 and −5. If it is very harmful it will score −5 but if it is only a slight problem, it will score −1.
   - If the resource has a **good effect** it will have a score between +1 and +5. If it is very good it will score +5 but if it is only just good it will score +1, of course.

2. Calculate the total score for the biogas scheme.

3. Find the score you gave to the thermal scheme.
   a. Compare it with the score for the biogas scheme.
   b. Which scheme has the higher score? This one is more sustainable than the other scheme. Explain why it is more sustainable.
   c. Is there anything that you think India could do to make the biogasifiers more sustainable?

4. You could find out about other energy resources such as hydro-electric power and nuclear power using the library or the internet, and use the environmental impact assessment form to give them a score too.

POPULATION AND RESOURCES

Study 3
Will there be enough resources?

The world's population keeps on growing, and our use of resources is increasing. Will there be enough resources to support people in the future? How can we change the way we use resources? These are some of the questions this study will look at.

3.1 How do we use our resources?

Resource use will continue to change. Geographers try to consider what changes there are likely to be in the future so that people can manage the resources to meet future needs.

In Chapter 5, Population Dynamics, on pages 102–117 we considered the ways in which the world population will change. Different areas of the world will have more or less people in years to come. Different areas of the world use different amounts of resources. **Figure 1** shows some of these differences.

Figure 2 gives information about world energy consumption. It shows the different amount of energy used per capita in different areas of the world. It again proves the idea that different areas of the world use different amounts of resources. It is the **less economically developed areas** that use less energy.

MEDC (North)	LEDC (South)
25% of world population	% of world population
Has 86% of industry	% of industry
Consumes 80% of world energy	% world energy
Uses 70% of world fossil fuels	% of world fossil fuels
Emits 60% of carbon dioxide	% of carbon dioxide
90% of cars	% of cars
Uses 350–1000 litres of water per person per day	Uses 20–40 litres of water per person per day

- The North consumes 15 times as much as the South
- 1 extra child in its life in an MEDC consumes the same as 30–50 children in an LEDC

25% of world population is starving
25% of world population is malnourished
1.3 billion live in absolute poverty
15% of the world's land is overgrazed and poorly farmed.

Figure 1 How fair is this? Different lifestyles in MEDCs and LEDCs

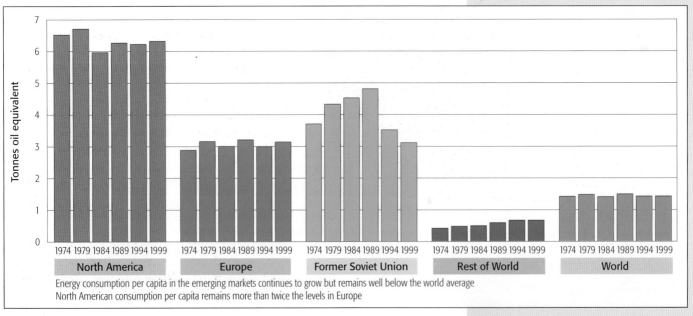

Figure 2 Energy consumption per capita throughout the world

Study 3 Will there be enough resources?

However, as these less-developed areas of the world develop more industry and use higher levels of technology (use more machinery) the demand for energy will grow. As well as this there are so many people in the LEDCs – when they start to use more energy the demand will rise enormously.

What changes can we expect in the future?

As well as predicting the change in demand, geographers also considered what they thought would happen. They wanted to work out whether we had enough resources left.

As long ago as 1798, Thomas Malthus wrote a book explaining his concerns for the future. He believed that population grew at an exponential (geometric) rate. That means that it doubled at each stage – 1:2:4:8:16. (On page 108 we saw how quickly families could increase.) Each stage was 2 times more, or bigger, than the one before. Malthus also believed that food production could only increase arithmetically. That means that it would only get bigger by adding one unit at each stage. People would add one unit at a time as more land was used, i.e. 1:2:3:4:5.

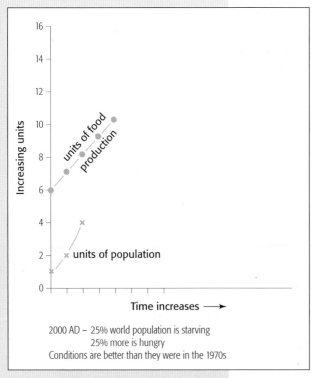

Figure 3 Malthus' prediction for population growth

Figure 4 Population pressure plus unexpected weather can add up to disaster: the Save the Children Fund tries to help

Over to you

1. Complete the statistics for the LEDC half of the chart in **Figure 1**. Use this to work out the figures:
   100 − MEDC % = LEDC %.

2. Study **Figure 2** and describe how world use of energy varies between different areas.
   a. Which areas use most?
   b. Which areas use least?
   c. What will happen as LEDC countries develop and use more machinery?

3. Look at **Figure 3**. It is part of a graph to show Malthus' prediction.
   a. On a copy of the graph complete it to show the changes Malthus suggested. Remember that population always doubles at each stage. Food production only gets bigger by adding one unit at each stage.
   b. Shade in the area where there would be too many people and not enough food. Where this happens people would die of starvation or famine. Label this area on your graph.
   c. Do you think that this is the situation in the UK today?
   d. Is it a true picture of other parts of the world? Look at the charity leaflet in **Figure 4** for ideas. There is also useful data alongside **Figures 1** and **3**.

3.2 What does the future look like?

People thought about Malthus' ideas (page 139) for a long time. The world managed to feed the population enough in most countries. In many MEDCs people had more than enough.

The Club of Rome

In 1972 a group of scientists, politicians, and university lecturers discussed the world situation and suggested a different viewpoint. They were called The Club of Rome and they thought that if population, industry, pollution and amount of food and resource use all increased, the world would suddenly face a disaster. Essential resources would run out or the world would become too polluted for people to survive. Their prediction is shown in **Figure 1**.

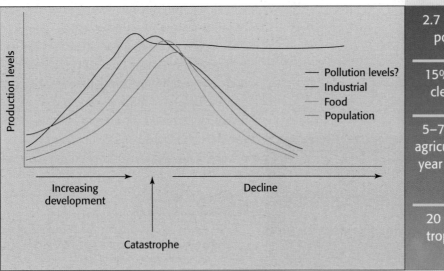

Figure 1 What the Club of Rome predicted

- 2.7 million die from air pollution each year
- 15% of world have no clean drinking water
- 5–7 million hectares of agricultural land lost each year to urbanization and degradation
- 20 million hectares of tropical rainforest lost each year

Boserup's view

Although many people listened to The Club of Rome, some people had a different opinion. They were more optimistic. Ester Boserup, a Danish economist, suggested that, as population grew, the great demand for food and resources would encourage scientists to develop new solutions to any problem of lack of supply. She thought that more food and resources would become available because of scientific developments.

Changes since the 1970s

Figure 2 shows some ways in which resource production and resource use have changed since the 1970s. Changes in energy resource use were also mentioned on pages 128–129.

People's opinions continued to change. A book published in 1984 considered the world as a living thing. The writers thought that people needed to look after the world better. Some of the things they said are shown in **Figure 3**.

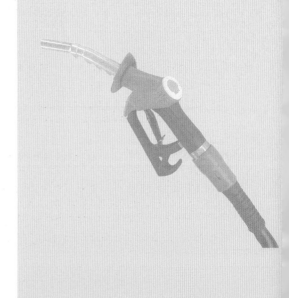

Study 3 Will there be enough resources?

Figure 2 Changing the ways we produce... and use... resources

Figure 3 Extracts from *The Gaia Atlas to Planet Management*

Messages for 'today's caretakers of tomorrow's world'.

At the moment we are surely the brightest and brainiest of the Earth's working parts. I trust us to have the will to keep going, and to maintain as best we can the life of the planet.

Lewis Thomas

We are making a start on the road towards sensible stewardship of our planet. While population grows and habitats continue to be degraded on every side, there has been an extraordinary outburst of awareness during the past few years. There are hundreds of environmental agencies at official levels, and at grass-roots level there are thousands of citizen groups; all are seeking to come to grips with our problems. Properly managed, the earth's more fertile lands and its forests could meet everyone's food and wood needs abundantly and indefinitely. Citizens of richer countries must ensure that their corporations and consumption patterns do not intensify land-use problems in the Third World.

Erik Eckholm, US environment expert.

Over to you

1. Look at **Figure 1**. It is the graph drawn by The Club of Rome.
   a. Describe what it shows.
   b. Do you think that it has influenced people's view of the future? Are their worries about pollution the same as your worries?

2. **Figure 2** shows some of the changes since the 1970s.
   a. Which changes have meant that more food could be grown? Make a list under the heading 'More food'.
   b. Which changes have meant that more resources are available? Make a list under the heading 'More resources'.
   c. Do you think that Boserup was right? Explain why.

3. Read the comments from the Gaia book in **Figure 3**. They are the thoughts of people in the mid-1980s. Write what they say in your own words.

4. You have been encouraged to write an essay for a competition. The title is 'When looking at the future, the future looks bleak.'
   - You can either agree or disagree with the title.
   - Use the information on pages 138–141 to provide evidence and ideas for your answer. (NB: People who think that the future looks bleak are afraid that bad events will happen.)

3.3 The way ahead – 1

In recent years the public has been involved in a number of ways to make resources last longer. Encouraged by local authorities and national governments, people have begun to reduce the amount of resources that are used up. They have begun to 'make a start on the road towards sensible **stewardship** of our planet'.

How can we achieve sustainability?

If we want to have a more sustainable approach to our use of resources we must 'use what the earth provides in a better way'. The changes that need to be made are:

- Reduce pollution of air, water and soil so that we can continue to grow food.
- Bring wasteland (in cities or countryside) back into use.
- Grow more food: by getting more from the land; by using more land; by growing plants that give a higher yield.
- Use fewer of the resources in more efficient machines.
- Waste less: don't throw so much away, make every bit of the resource useful, use the resource again and again.

In the information on these two pages, and the following two, you can see some of the ways that different people have responded. A few ideas are given here.

- Designers have tried to find alternative forms of transport.
- Governments and private companies have tried to encourage people to use public transport.
- Local councils have allocated bus and cycle lanes.
- House builders have incorporated more energy-efficient materials.
- Power companies have looked at more renewable sources of energy.
- Clothing manufacturers have turned from synthetic to natural fibres.

Figure 1 More people will cycle when safe cycleways are built

Figure 2 'Proud to be green': banks and businesses lead by example

Putting our own house in order

At The Co-operative Bank, we're not just urging our customers to 'think green', we're working to reduce the environmental impact of our own activities.

And our latest Partnership Report shows we are making real progress.

For example, thanks largely to our own use of energy from renewable sources, our net carbon dioxide emissions have fallen by 32% since 1997. And we're proud of being the UK's largest user of 'green' electricity.

If you'd like to know more about the bank's own commitment to ecological sustainability, please read our Partnership Report at www.co-operativebank.co.uk

Study 3   Will there be enough resources?

WHAT TO RECYCLE

Paper

About one-quarter of all household waste is newspaper and magazines.

These can easily be recycled into fresh newsprint at an economic price, so reducing the need for raw material.

Cans

The production of aluminium and tin cans uses up the earth's finite resources. However, metal may readily be recycled and turned into new cans using a fraction of the energy that would be required if raw materials were used.

Glass

Glass is difficult to dispose of and can leave dangerous shards when buried in the ground. Bottles, however, can readily be recycled into new glass containers many times over.

Clothes

The Council has arranged for charities to provide textile recycling bins around the Borough. Clothes can sometimes be cleaned and sold or alternatively the material can be re-used. Almost everything is recyclable.

Plastics

A recent innovation has been the introduction of plastics recycling facilities, initially at three locations. Most plastic bottles are acceptable but please do not deposit motor or vegetable oil bottles or any bottle caps.

Kitchen and garden waste

Kitchen and garden waste can also readily be recycled to provide plant food.

Other measures

Other recycling measures operating within the Borough include special collections of materials from sheltered housing schemes, council housing estates and council offices. The recycling of cardboard from the markets and other locations is currently being trialled.

GARDEN REFUSE

Almost all garden refuse can easily be recycled into rich compost. The Council supports this by making available composting bins at specially subsidized prices, from £12.50 each (normal retail price up to £50). Full details are available from the One Stop Shops or by telephoning the Environmental Help Line on 01992 785577 (minicom 01992 632150).

WASTE REDUCTION

Recycling is essential. However, it is even better to avoid creating waste in the first place. Here are a few ways of going about this:

▼ Take your own bag when you go shopping and avoid using carrier bags that you do not require

▼ Avoid using disposable items such as paper cups, plates and disposable razors

▼ Choose goods with less packaging – for example, select loose fruit and vegetables rather than pre-packed, and choose concentrated detergents and refills that use less packaging

▼ The government now requires manufacturers to reduce the amount of packaging they use but as consumers we can all bring extra pressure to bear

Figure 3 People need to learn how to recycle: the Borough of Broxbourne explains how in this leaflet

POPULATION AND RESOURCES

3.4 The way ahead – 2

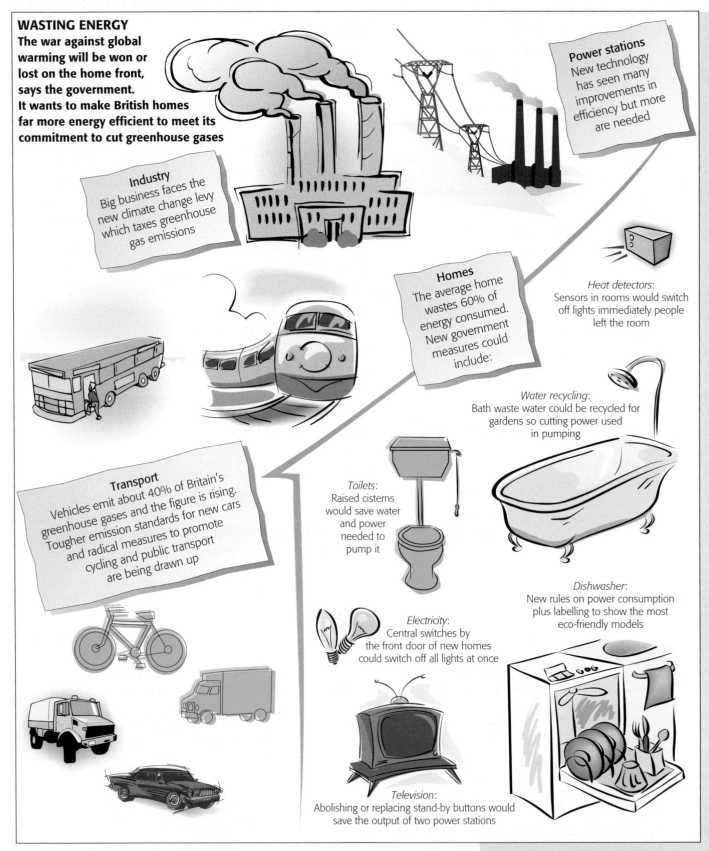

Figure 4 Energy: how not to waste it

Study 3 Will there be enough resources?

Policy maker	Strategy	Response hoped for	Your family response
Government	Grants for home improvement for energy efficiency	People insulate houses More efficient boilers in homes	
	Tax on petrol to reduce fuel use	People use cars less	
	Targets for recycling for local councils	Local councils respond to **Local Agenda 21** scheme	
	Laws on pollution reduction – smoke-free zones	Power stations and individuals burn less polluting materials	
Local authority	Promote recycling by leaflets	People collect and **recycle** more items	
	Introduce electric vehicles	Reduce energy consumption	
Companies	Investigate renewables for electricity production	Change to renewable energy	
	Use less packaging (Sainsbury, Waitrose)	People buy re-useable shopping bags	
	Make clothes with natural fibres	People buy cotton and wool clothes	
Schools	Set up walking buses	Children walk to school in 'bus'	
TV shows	Inform public	People make environmental choices	
	Organize collections	Blue Peter viewers collect recyclables!	
Farmer	Introduce high yielding varieties (IR8 rice)	Increased yields	
	Test GM crops	Introduce GM crops	
	Introduce organic crops	Use fewer non-renewable chemicals – fertilizers, biological controls	
Conservationists	Warn of future disasters – global warming, acid rain	Greater co-operation to reduce CO_2 emissions etc.	
Community worker	Charity shops	Recycle clothes, household goods	
Planners	Allow development on new areas of land (reclaim marshland)	More land available for farms and businesses	
	Land reclamation (**brownfield sites**)	Improve land that has become unproductive	

Figure 5 Changing the way we use resources: who's involved?

Over to you

1. Look at **Figures 1–4** on pages 142–144. Study **Figure 5**. Read what they all say and make some notes ready for the discussion in question 2.
2. In groups of 4, discuss what the following are doing to encourage us to develop a more sustainable lifestyle.
    - Government
    - Local authorities
    - Large companies
    - Conservationists
    - Community helpers
    - The media
    - Ordinary people

    Think about grants, taxes, publicity, changing work practices (e.g. tele-working), school campaigns, charity shops.
3. What do you do to save resources?
4. What else would you like to do?
5. Can you and your class, or friends, arrange for these ideas to happen? Don't write an answer for this – just do it!!

POPULATION AND RESOURCES

3.5 Looking back at China

We have looked at some of the ways that people are moving towards a more sustainable future. Countries in the developed world already have a high standard of living and people have enough wealth to be able to make these changes to the way they live without starving or going without income. For other countries this is more difficult.

In less economically developed countries the governments have a hard task. They have to provide for more and more people. These people may be very poor and may have had little education and not learnt about the changes that are needed. They may have lived in one small village all their life, and may be unaware why their decisions are important to the whole world.

The governments must educate their people, improve the methods that they use, keep providing enough food and jobs for all the people and still try to have a more sustainable approach.

Near the beginning of the chapter on population dynamics, on pages 104–105, we read a story about the Huang family. We learnt that their baby son had been killed. It was very difficult for us to understand why this should happen. Perhaps now we are beginning to understand the big problem for the world. Perhaps we can understand how difficult it is for the Chinese.

The newspaper extracts in **Figures 1** to **3** tell of some other ways that the Chinese government is trying to cope with its huge population and the use of its resources.

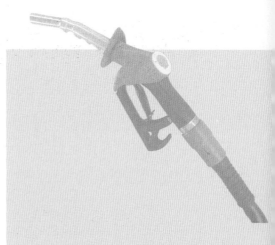

Figure 1 The Chinese: doing their best with what they have

Farming change in China

EACH family has a plot of land and can grow whatever it wishes. Some food must be grown and sold to the state. Most is then used to feed the family and animals but the rest can be sold at market.

The farm size varies. If a family has a baby it will be given land from a family whose child has just left home.

There are few tractors and machines. Sometimes the man pulls the plough and his wife guides it. There is very little spare land. Even motorway verges are farmed. A wide variety of fruit and vegetables are grown. Pigs, chickens and ducks are raised in the village. There are few cows and the Chinese don't really like to eat cheese or drink milk (cows eat too much food).

Soil erosion is a problem and much stored food rots away before it is eaten. Sometimes the government gives people machinery, but tractors often go rusty because no one is trained to use them.

Too many mouths to feed

One-fifth of the world's population lives in China and feeding these people with only 7% of the world's agricultural land keeps farming at the top of government's agenda. Even with the one child policy and population growth slowed to just over 1% per annum, China simply has too many mouths to feed – there are more people learning English in China today than there are native English speakers on the whole planet!

In the last year the Chinese government has begun to admit that even with improved storage, research into new varieties and better farming methods, it will probably be unable to achieve even the basic goals of producing as much wheat and rice as it needs in the long term.

Consumption keeps increasing ahead of production and the Chinese are not debating the pros and cons of GM crops. They accept them as a necessary part of their future.

Figure 2 Population: a problem that keeps on growing

Study 3 Will there be enough resources?

China moves mountains for clean air

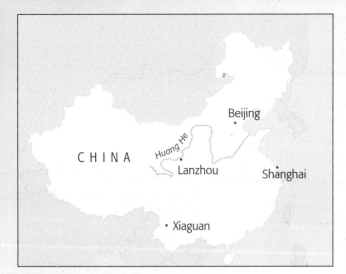

LANZHOU, the Chinese city that in 1998 won the dubious accolade of being named the world's most polluted place, is toying with a radical solution: demolishing a mountain to let in some fresh air.

Lanzhou lies in a narrow valley, hemmed in by hills. It stretches for 25 miles and in places is as narrow as 1½ miles. Pollution from hundreds of chimneys produces a choking mixture.

The city is in the middle of a dry desert region in the north western province of Gansu. The air is dusty and the elevation of the city, which lies 5000ft above sea level, makes the fumes from coal furnaces dissipate only slowly.

Removing the hill would unplug the bottle and allow pollution to escape, supporters of the plan say. Its critics say that this is nonsense. 'Lanzhou is surrounded by hundreds of hills and mountains. How will removing one solve the pollution problem?' Wang Xiumin, a foreign trade official, told the *South China Morning Post*. 'It's a crazy idea.'

Lanzhou is also trying more traditional methods: planting trees to absorb carbon dioxide, converting coal-fired furnaces to gas and closing the dirtiest factories. That has had some effect, although the city still has to endure 20 weeks of highly polluted air every year. Many people resort to wearing smog masks.

'In the winter it is even harder to breathe than usual because people use coal for heating,' said Chen Jun, a taxi driver. Like other cabbies he is restricted by pollution-control regulations to working on alternate days.

The Big Green Hill that dominates Lanzhou. *City officials hope that removing the summit will allow pollution to escape from the choking valley*

Figure 3 Getting some fresh air... how far will you go?!

Over to you

1. Read the newspaper articles about China, its people and resources.
2. Make a list of the ways that the government is helping to match population and resources.
3. Explain what it is doing to try to reduce pollution caused by industry using fossil fuels.
4. Write a letter to Huang Quisheng's family (pages 104–105) and explain why the Chinese government has to be so strict about the number of people born

    Remember there are already over a billion people in China.

    Every child will need food, education, a job, a good standard of living and care in old age. Perhaps you could start by explaining that China is one part of the world, and the whole world must face the future and its problems together.

POPULATION AND RESOURCES

3.6 Population and resources – whose responsibility is it?

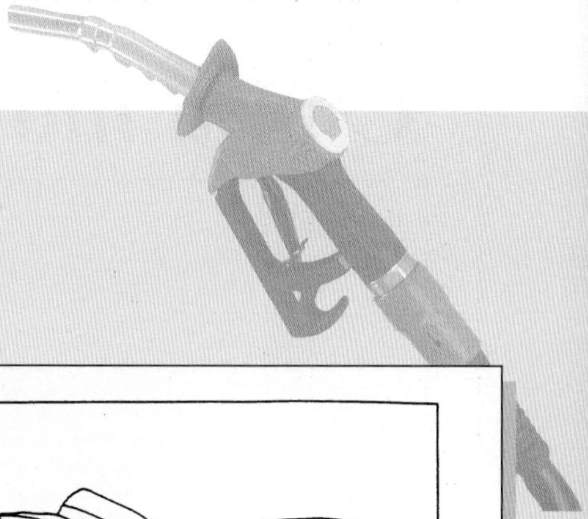

Throughout the chapters on Population Dynamics and Population and Resources, we have thought about changes and the future. It is your future and your world. What will it be like and what should you do?

The exam unit is called 'Providing for Population Change'. We have studied population sizes and then resources. The cartoon, **Figure 1**, sums up some people's attitudes to the issue of providing for an increasing world population.

When you have answered the questions in 'Over to you' you will have used the ideas you have learnt to make comments about the future, your future. If you understand all these ideas all you will need for the exam is a good detailed knowledge of the case studies.

■ Case studies

1. Population change in an MEDC – we looked at Germany.
2. Population change in an LEDC – we looked at Malawi.
3. A thermal energy scheme in an MEDC – we focused on France
4. A renewable energy scheme in an LEDC – the example was biogas in India.

Figure 1 Who should do what? A cartoon from *New Internationalist*

Study 3 Will there be enough resources?

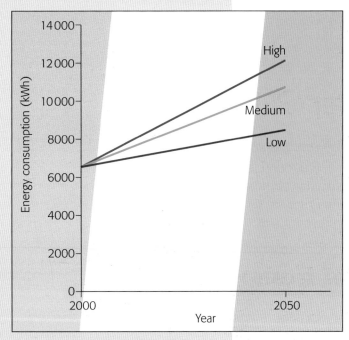

Figure 2 The predicted growth in electricity consumption in France from 2000 to 2050

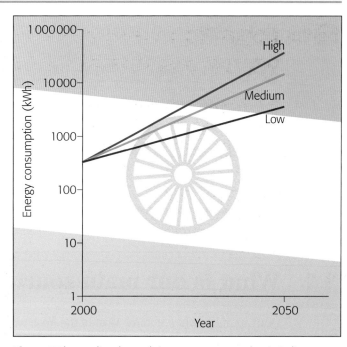

Figure 3 The predicted growth in energy consumption in India from 2000 to 2050

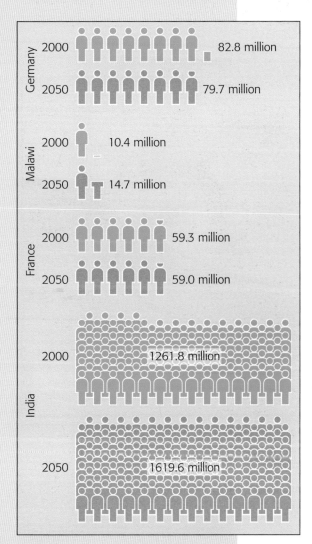

Figure 4 Predicted population growth in some MEDCs and LEDCs

Predictions for the future

The graphs in **Figures 2** and **3** show predictions in the growth of electricity and energy consumption in France and India for the future (2050). The pictograms in **Figure 4** show the predicted number of people in each country we have studied in the population chapters. As no one can be certain of the future, geographers and analysts give different 'scenarios'. They take the present or expected pattern and extrapolate it into the future. They then calculate a possible but higher scenario. They also calculate the most likely lowest scenario. In this way planners can predict for the future. Notice the scale for India in **Figure 3**. It is a logarithmic scale because the changes are so big.

Over to you

1 Using the work that you have done in this chapter, produce a response to the cartoon, **Figure 1**. It could be as a letter, another cartoon strip or an essay. You might want to refer back to Chapter 5, Population Dynamics. You will need to discuss:

   a What is being done about population increase in the world. You will probably want to mention LEDCs and MEDCs separately. (See **Figure 4**)

   b What is 'wrong' with people? Why could they be seen as a burden? Why are they an asset (good) for a country?

   c How do we cope with growing numbers of elderly people?

   d How do we cope with growing numbers of young people?

   e Why is it also a problem of resources? Look at **Figures 2** and **3** as a reminder.

   f What are people doing about resource consumption (use).

WATER

Study 1
Why so much fuss about water?

Fresh water is a scarce **resource**. It is found on the earth's surface in rivers, lakes and reservoirs (man-made lakes) and underground in **aquifers** (rocks capable of storing water). The importance of these different stores varies from place to place depending primarily on both climate and geology, but also on our ability to access them.

1.1 What is our main source of fresh water?

Don't grumble when it rains. Most of the fresh water on which life depends comes from precipitation, mainly in the form of rain and drizzle (only 6% of global precipitation falls as snow). Nearly 90% of the moisture (water vapour) in our atmosphere is evaporated from the oceans. It is this circulation of water, known as the **water cycle**, which maintains the supply of fresh water.

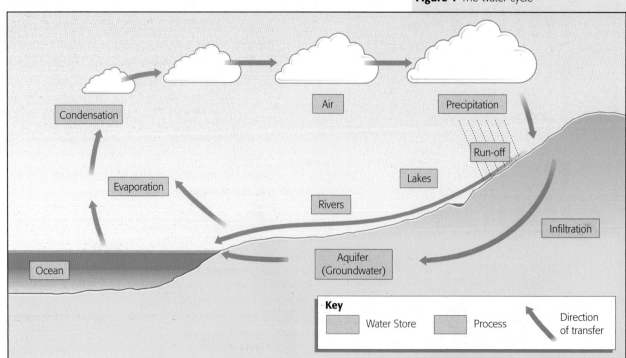

Figure 1 The water cycle

■ Where are the water stores in the United Kingdom?

In Highland Britain water is stored mainly in surface stores. In Lowland Britain much water is obtained from aquifers in **sedimentary** rocks. **Figure 2** shows how Yorkshire obtains its water.

Study 1 Why so much fuss about water?

Figure 2 Yorkshire Water supplies and major demand areas

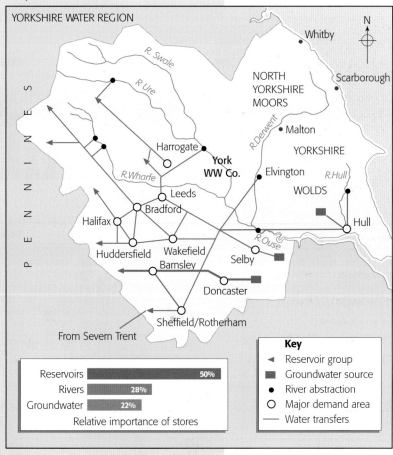

Figure 3 Yorkshire Water stores

Reservoirs
- Manmade lakes formed by constructing earth or concrete dams across valleys
- Most are in the upper catchment of the Pennine rivers
  - Rain and snow recharge the reservoirs during winter
  - Several streams flow into a reservoir
  - Most reservoirs are small and hold only 120 days water supply

Rivers
Main suppliers are the Wharfe, Ouse, Derwent and Hull

Derwent receives many small tributaries from the North Yorkshire Moors

Groundwater supplies
Major aquifers – Chalk of the Wolds and Corallian Limestone of the North Yorkshire Moors

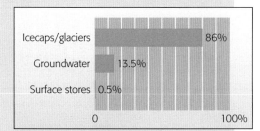

Figure 4 Global freshwater stores

Figure 5 Groundwater

Groundwater
- most of the rain falling on the earth's surface makes its way underground
- accounts for 94% of accessible freshwater
- 40% of groundwater is within 1 km of the surface of the ground
- is a major source of water in arid regions because of its reliability
- often the largest store of water in a drainage basin
- is usually safe water

Over to you

1 a Using **Figure 4** explain why much of the world's fresh water is of little use to people.

  b Use your atlas to find the Columbia, Rhine, Indus, Ganges and Brahamaputra rivers. Name either the glaciers which feed them or the mountain ranges in which the glaciers lie.

  c Where are the world's two largest ice sheets?

2 Although the oceans store 97% of global water, we cannot use it directly as a source of drinking water. Why not? Find out by which process the oceans can be made useful and why so few countries obtain water this way.

3 a Use a CD ROM or atlas map to name some of the major reservoirs in England and Wales. Why are they located in highland regions?

  b Now find a map showing the distribution of population in England and Wales. Write a sentence summarizing water supply and demand in England and Wales.

4 Using the information in **Figure 5** explain fully why groundwater is such an important source of water.

5 Suggest why water is abstracted from Elvington (**Figure 2**) and not from the upper reaches of the Derwent. Use **Figure 2** on page 92 to help you.

WATER

1.2 Why do water supplies vary?

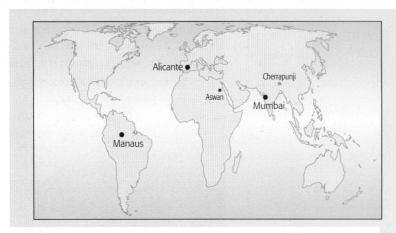

Rainfall varies from place to place

Imagine living in either of these two places: Cherrapunji (Assam, India), which receives over 11 000 mm in a year, or Aswan (Egypt), where it hardly ever rains. Dramatic differences. But these extremes illustrate how greatly amounts of rainfall can differ globally.

You may know that the wettest parts of the world are the areas which experience equatorial and monsoon climates, but it may surprise you to learn that the polar regions are among the driest areas. The world's largest arid area stretches from the Middle East across North Africa to the Atlantic Ocean. This area experiences more or less permanent drought. Rainfall is not only low, but very unreliable. The occasional rain event is short-lived. In a matter of hours heavy downpours turn dry river beds into raging torrents for just a brief period.

Rainfall varies over time

Not only do amounts of rainfall vary, but so does the seasonal pattern of rainfall. In regions in India which have a monsoon climate, and in the Mediterranean, wet season runoff from heavy rains results in flooding. Much water is lost to the sea and wasted. But in the dry season rivers are reduced to a mere trickle, just when water is most needed.

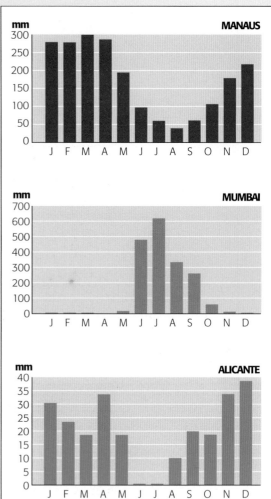

Figure 1 Rainfall graphs for Manaus, Mumbai and Alicante

Figure 2 Now you see it... the contrast between the dry and the wet season in Indian rivers.

Study 1 Why so much fuss about water?

Ground water stores are also affected by rainfall. In the Yorkshire Wolds the **water table** is at its highest in early to late spring, but falls by 10 m to its autumn low.

What the rainfall graphs in **Figure 1** do not show is that rainfall totals vary from year to year and within countries. Nor is drought confined to the climates already mentioned. The UK experiences occasional drought. That of 1995–97 was severe. Yorkshire Water was seriously affected. Not only were river flows reduced but reservoirs fell to danger levels and water had to be brought in by tanker from Northumbria.

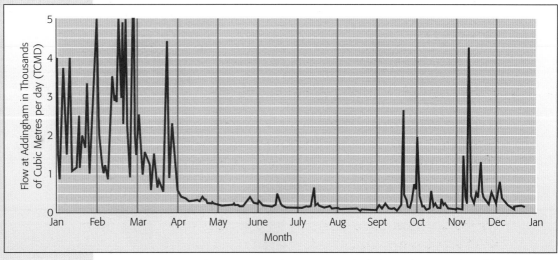

Figure 3 Hydrograph of the River Wharfe, in Yorkshire, in 1995

Figure 4 Rainfall tables: two examples from Yorkshire

Rainfall in millimetres Great Walden Edge (Yorkshire Pennines)															
	Oct	Nov	Dec	Jan	Feb	Mar	Apr	May	June	July	Aug	Sept	Oct	Nov	Dec
1994–5	161.9	188.9	272.2	242.9	220	186.6	43.2	55.1	31.2	64.6	18	107.8	69.4	53.2	61.7
Long-term average	133	140	139	129	91	109	91	88	91	91	117	124	133	140	139

Cottingham (East Yorkshire)															
	Oct	Nov	Dec	Jan	Feb	Mar	Apr	May	June	July	Aug	Sept	Oct	Nov	Dec
1994–5	44.3	63.5	50.7	97.5	68.7	57.8	25.9	34.1	42.5	27.5	7.5	109	16.9	52	68.6
Long-term average	53	64	64	55	41	53	50	53	52	49	66	54	53	64	64

Over to you

1. Users of water need a reliable source of water. For a named country, explain the problems caused by an unreliable water supply.

2. The water year in the UK runs from October 1st–September 30th. Using **Figure 4** calculate the total rainfall for Great Walden Edge and Cottingham in Yorkshire for the water year 1994–5, and for the long-term average. Suggest why the long-term amounts are different.

3. For one of the rainfall stations, draw a bar graph showing the long-term average. Then draw a line graph over this for the 1994–5 rainfall. This will show the 1995 drought. Now compare the rainfall amounts.

4. Explain why the main period of reservoir and **aquifer recharge** is during the winter 6 months. Reservoirs in the Pennines only hold enough water for 120 days' supply. Why were water stocks very low by the end of summer 1995? Why were domestic customers of Yorkshire Water being asked in January 1996 to use beakers of water when cleaning their teeth instead of a running tap?

1.3 Is there enough water for everybody?

■ Which countries were the lucky ones in 2000?

Rainfall pattern and total rainfall are not the only factors affecting how adequate a country's water resources are. Some countries are very densely populated. Other countries have small populations.

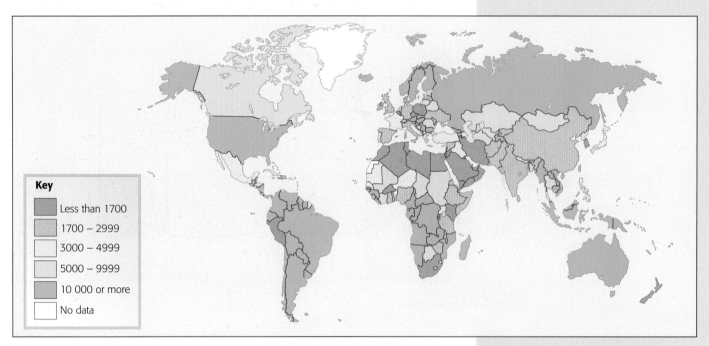

Figure 1 Freshwater resources per head of population in cubic metres (m^3)

Key
- Less than 1700
- 1700 – 2999
- 3000 – 4999
- 5000 – 9999
- 10 000 or more
- No data

By the year 2000 approximately one-third of the world's population experienced **water stress**. (Water stress is defined as when the amount of water consumption is more than 10% of the renewable freshwater supply.) **Figure 2** shows the average water needs per person per year. However, in LEDCs south of the Sahara, the amount of water used for domestic purposes in a year is only 5 cubic metres. When water is not on tap but involves walking long distances to access, then the amounts people use are small.

■ And who will have enough water in 2025?

On 12th October 1999 the United Nations Secretary General declared that the world's population had reached 6 billion. Global population increased threefold during the twentieth century, but water consumption increased sixfold over the same period. More significantly, the rate of increase in consumption is accelerating. The highest rates of population growth are in those parts of the world where water is already in short supply, and it is these areas where demand is set to rise. By 2025 at present consumption rates, two-thirds of the world's population will be experiencing water stress. **Figure 3** shows how different users will compete for water in LEDCs.

Figure 2 Water needs per year

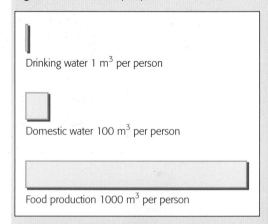

Drinking water 1 m^3 per person

Domestic water 100 m^3 per person

Food production 1000 m^3 per person

Study 1 Why so much fuss about water?

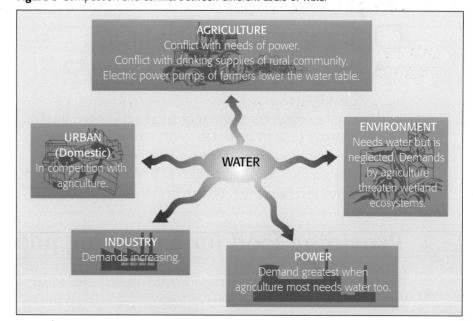

Figure 3 Competition and conflict between different users of water

Figure 4 Comparing water resources and economic development

Country	Freshwater resources		GNP $ 1998 World Bank method of calculation	Rank (out of 206) of economy
	Available per capita in 1998 in m³	Annual use as percentage of total 1990–8		
Brazil	42 459	0.5	4630	68
Canada	92 142	1.6	19 170	26
Egypt	949	94.5	1290	121
Israel	184	155.5	16 180	32
Norway	88 673	0.5	34 310	4
United Kingdom	2489	6.4	21 410	22

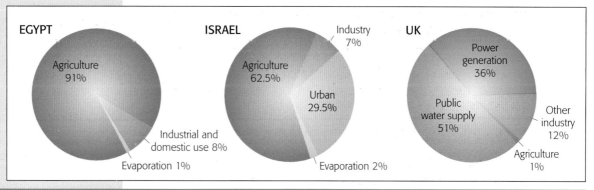

Figure 5 Contrasting water usage in Egypt, Israel and the UK

EGYPT: Agriculture 91%, Industrial and domestic use 8%, Evaporation 1%

ISRAEL: Industry 7%, Agriculture 62.5%, Urban 29.5%, Evaporation 2%

UK: Power generation 36%, Public water supply 51%, Other industry 12%, Agriculture 1%

Over to you

1 Look carefully at **Figure 1** and a map showing the global annual precipitation.
   a Name two countries in **Figure 1** which do not fall into the category for the amount of water per capita you would have expected.
   b Give reasons for your choice.

2 Use **Figure 4** to name: **a** a water-stressed LEDC; **b** a water-stressed MEDC; **c** a water-rich LEDC; **d** a water-rich MEDC.

3 Using **Figure 5** suggest some reasons as to why the proportion of water for domestic, agricultural and industrial uses varies from country to country.

WATER

Study 2
Demand is rising. What can we do?

Worldwide, the largest consumer of water is agriculture. It takes 1000 m³ of water to provide the food you consume each year. The world population will increase by 3 billion in the next 60 years, making a total world population of 9 billion. The impact of this predicted increase is not difficult to calculate.

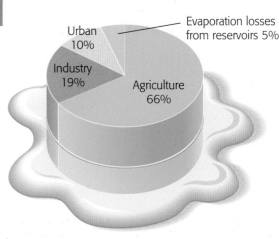

Figure 1 Water use throughout the world

2.1 Providing food for a growing population

The FAO (Food and Agricultural Organization) estimates that about 60% of the extra food we shall need will come from irrigated farming. Water demand for this purpose will increase by at least 17% and perhaps by 50%. Careful management of water will be critical.

■ Egypt's food gap

By the 1950s food production in Egypt could no longer keep up with the rapid population growth. There was a food gap. And there was a shortage of irrigated land. Only 4% of the country can be cultivated: mainly the Nile delta and a narrow strip of land, about 8 km wide, either side of the River Nile. These are areas which not only have fertile alluvial soil but also, importantly, can be easily irrigated. Without **irrigation** there can be no farming as with little or no rain there is a lack of soil moisture. Most farmland is irrigated by flood irrigation. Soils are saturated prior to seeding and more water is applied at intervals during the growing season. This irrigation makes heavy demands on water: 90% of Egypt's allocation of Nile water. But the Nile is virtually the only source of water in Egypt – so just consider the implications of future increases in demand.

Water - the Key Facts

- The bulk of water used in agriculture is used in irrigation
- Irrigation makes indirect demands on water as electricity is needed for pumping
- Irrigated farming produces over 40% of the world's food on 16% of the globe's cropped land
- Irrigation provides essential water for crops in countries where soils lack moisture due to permanent or seasonal drought
- In India 60% of the food produced is dependent on irrigation
- The highest rates of population growth are in Africa and Asia
- Large regions of Asia and Africa already experience water shortages
- 2.4 billion people in LEDCs depend directly on irrigated agriculture for food and livelihood

Figure 2 The key facts about water use in agriculture

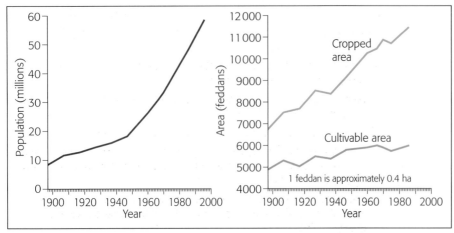

Figure 3 Population growth and cultivated land in Egypt

Study 2 Demand is rising. What can we do?

■ Irrigation can increase the area under cultivation

Figure 4 Cultivated land in the Nile Valley

Egypt wanted to step up its agricultural production to bridge the gap. A major scheme of land reclamation was necessary. Areas of land were to be reclaimed from the desert edges, always close to land already under irrigation. In all, it would increase the area under irrigation by half as much again.

By 1997 population growth had again put pressure on land and food resources. But Egypt had used all its easily cultivated land so President Mubarak announced plans for an even more ambitious scheme. By the year 2002 one-tenth of the annual flow of the Nile will be taken by a canal to the New Valley Province in the Western Desert. The irrigated farmland will grow not only wheat and other food crops for Egypt but also high-value crops for export.

There are also plans to irrigate large areas of land near the Sudanese border, using groundwater, and in Sinai, using water from the Nile.

■ Irrigation can increase crop yields

With the completion of the High Aswan Dam and an assured supply of water, no longer is crop failure a hazard and three harvests a year are possible. Also, where new farming methods are used and water is applied to crops at the right time and in the correct amounts, crop productivity can be increased, especially where new high-yielding varieties are sown.

Over to you

1. Draw a sketch of **Figure 4**. Annotate your sketch to show why the valley floor is so important for farming.
2. Look at **Figure 3**. Explain why the area of cropped land is greater than the cultivable land.
3. Look at **Figure 3** and explain why there was a food gap.
4. List the ways in which Egypt is bridging its food gap. How does this tie in with the issue of providing food in the future? Is there any other way Egypt could access food?
5. Is it in fact sensible for a water-short country to use so much water for food? What other demands do you think will be made on water as Egypt develops?

WATER

2.2 How are urban areas in LEDCs coping with water demand?

Approximately half the world's population now lives in urban areas. By 2025 this proportion will have increased and nearly 4 billion people will live in urban areas.

	Percentage of the world's urban population		Number of the world's 100 largest cities	
	1950	1990	1950	1990
Africa	4.5	8.8	3	7
Asia	32.0	44.5	33	44
Europe	38.8	22.8	36	20
N. America	14.4	9.2	18	13
Latin America	9.3	13.8	8	14

Figure 1 The world's urban population

- Since the 1950s the greatest urban growth has been in LEDCs and this will continue because their national populations have grown rapidly and will continue to do so.
- Rural–urban migration will speed up, so increasing urban populations.
- As industrialization continues to grow, demands for water will greatly increase.

Water demand in urban China

At present only 30% of China's population is urban. This is only a small proportion of the total population, but is still a large number of people: 400 million! But the urban population will increase more rapidly now owing to the very fast economic development in China. Urban demand for water has been rising by over 6% a year since 1990. Water for urban use will compete with the needs of agriculture in the surrounding areas. China needs to develop more water resources to satisfy current urban demands – of which only approximately two-thirds are being met. But equally important is the need to treat urban waste water.

Figure 2 A McDonald's in Shanghai: bright lights, big city... and water resources under pressure

158

Study 2 Demand is rising. What can we do?

■ Shanghai

The population of Shanghai, China's largest city, is growing rapidly. It has doubled in four decades. Nearly 8 million people live in the urban core, 13 million in greater Shanghai, and there is a migrant population of 4–5 million. Demand for water is rising because the population is increasing and because for many people living standards are getting better. Overcrowded tenements where there was no indoor tap are being pulled down and residents are moving into new apartments on the western edge of the city, or across the Huangpo into the new area of Pudong. No longer do they have to share one tap between 100 households: each flat has its own water supply and indoor sanitation. As more and more people get access to piped water, demand will accelerate. And Shanghai is a major industrial city, and industrialization and its overall economic growth will also increase its demand for water.

Shanghai's water resources are already overstretched. The Huangpo River supplies most of its water for domestic, industrial and agricultural purposes. Groundwater abstraction is causing subsidence.

Shanghai is looking further west for new sources. Not only are there insufficient supplies but water resources in the area are affected by pollution. All drinking water in Shanghai must be boiled. But Shanghai is not the only rapidly growing city of the Yangtze delta. The whole area is highly urbanized. On the outskirts of Suzhou, only 86 km to the west, a new industrial town is being built which will add to the problem of water supply.

Figure 3 The new area of Pudong in Shanghai

Figure 4 Many new residents of Pudong will live in developments such as Chrysanthemum Park, shown here

Method of supply	Percentage of population supplied	Comments
Piped water	50%	Not necessarily safe nor continuously available
Yard taps, public standpipes hand-pumped	25%	
Water vendor	25%	Price often 12 times higher than piped supply

Figure 5 Urban water supply in LEDCs in 1991

Over to you

1. Why is demand for water in LEDCs increasing? Why is demand set to rise even further?
2. Draw a diagram to show the knock-on effects of urbanization on water demand.
3. Give two reasons why waste-water treatment is a key issue in increasing water supply.

WATER

2.3 What's happening in MEDCs?

In South-West USA urbanization still continues. The spread of urban areas in California devours 80 000 ha of land a year. Los Angeles is the second largest metropolitan area in the USA. Its population has now reached 14.5 million. Supplying water to meet this rapidly growing city has always been a problem. As early as 1905 work started on a 373 km aqueduct to import water from the Owens Valley, much to the annoyance of the farmers whose livelihoods were affected. And since the mid-1930s additional water has been transferred by aqueduct from the River Colorado over 300 km away.

■ What is the water used for?

Other cities are also growing in the South-West, such as Phoenix and Tucson in Arizona. But Las Vegas, in Nevada, is the fastest growing of them all. With 65 000 new inhabitants a year it has now reached a population of 1 million. Supplying water to meet residential demands in all these cities is a key issue. In Tucson, 470 litres is needed per person a day. Enjoying high standards of living means a swimming pool and water for trees and lawns. But this, in an arid region with only 100 mm rain a year and summer temperatures of over 25°C, is a very wasteful practice. Up-market residential areas have been built around artificial lakes, although such developments are no longer permitted.

Leisure and tourism also use plenty of water. Golf courses are major consumers of water. On the positive side, some hotels now irrigate their gardens with recycled waste water. Las Vegas is a major tourist centre and in 1999 received 33 million visitors, all needing vast quantities of water.

Industry also needs water, but in these three cities this consists of either light industry or services, making small demands on water.

Figure 1 Los Angeles: population 14.5 million ... and counting

Figure 2 This residential area in Los Angeles shows the reason for high water usage

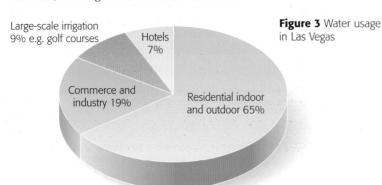

Figure 3 Water usage in Las Vegas

Study 2 Demand is rising. What can we do?

■ Where does the water come from?

Where does all the water come from in this desert environment? All three cities, Phoenix, Tucson and Las Vegas, like Los Angeles, are now heavily dependent on the River Colorado to meet their rising demands. Phoenix and Tucson access the Colorado's water from Lake Havasu by means of the Central Arizona Project (CAP). Las Vegas draws 82% of its water from Lake Mead impounded by the Hoover Dam, the first to be built on the Colorado. But demands are accelerating and Las Vegas is seeking new supplies in Eastern Nevada. This will be financially and environmentally expensive to develop.

Nor does the Colorado just supply water. Its hydro power is vital for homes, industry, agriculture, and tourism. Imagine the amount consumed by hotels in Las Vegas!

Figure 4 The Strip, Las Vegas, at night

● Sport causes water conflict

LEISURE has become another consumer of water. In the UK new golf courses replace farmland. In western USA golf courses are also the latest craze. But in this arid part of the States water supplies are already under stress. The amount of water abstracted to water greens is causing environmentalists concern.

Falling river levels affect water habitats whilst felling trees to make way for this space-consuming activity increases silt loads and so reduces water quality. And much to the surprise of visitors is the golf course at Furnace Creek in Death Valley with a rainfall of only 42 mm. With temperatures reaching over 37°C from May through to October, water consumption does not bear thinking about.

Nor is this the end of it. At the interface of desert and delta and in the Red Sea resorts of Egypt, new golf courses are another means of luring the golf enthusiast from less sunny climes and providing a welcome break for Cairo's business clientele.

The good news is

Figure 5 Demands on water from sport and leisure

Over to you

1 In North America and Europe increases in industrial use of water will only be small, unlike in LEDCs. In some industrial cities industrial demand has actually declined. Explain why (think about the way industry is changing).

2 Urbanization leads to higher demands for water not only for domestic needs but in other sectors. Draw a diagram to show these knock-on effects.

3 In the UK domestic demands are rising for both internal and external domestic uses. Make a list of activities which consume large quantities of water.

4 Conflicts are arising between different water-using activities in MEDCs. The increasing demands for leisure and the needs of the environment is one example. Write the next paragraph of the magazine article in **Figure 5** discussing non-polluting water-using activities.

2.4 How has Egypt increased its water supply?

	Jan	Feb	March	April	May	June	July	Aug	Sept	Oct	Nov	Dec
Entebbe	94	96	168	281	257	98	64	84	83	112	157	122
Addis Ababa	13	38	67	86	86	137	279	299	191	20	15	5

Figure 1 Rainfall in mm

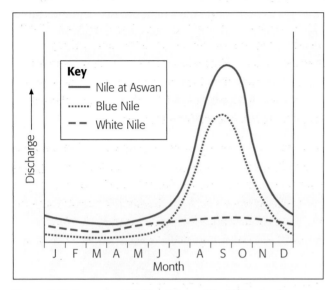

Figure 2 Hydrograph of traditional flow of the Nile

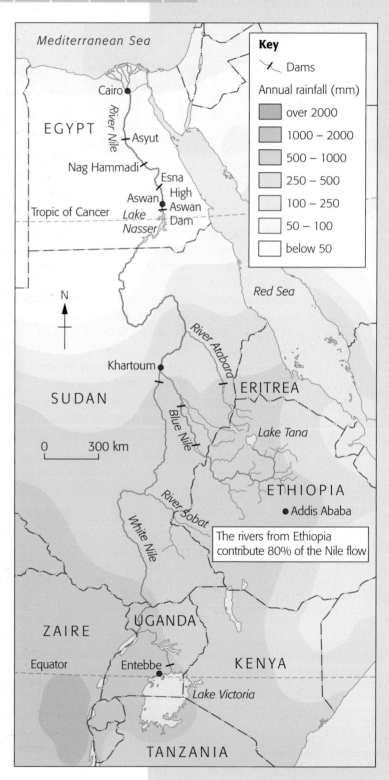

Figure 3 The Nile Basin

The River Nile floods annually. The height of the flood level varies enormously from year to year. Until the twentieth century, excessively high water caused widespread devastation in the Nile valley. Low flood levels meant that there was insufficient water for crop growth. Several successive years of low water resulted in famine. So, in 1902 a dam was built across the Nile at Aswan to help control the flow of the river and store some of the flood water which could then be released during the season of low water. But this dam could not store the full flow and much-needed water was lost to sea and wasted.

Study 2 Demand is rising. What can we do?

Why was it necessary to build the High Aswan Dam?

Most of Egypt receives less than 125 mm of rain a year and the Nile is the only river in Egypt. So Egypt is almost totally dependent on the Nile for all its water. The management of the Nile is vital for all economic activity, but especially agriculture. With high temperatures and evaporation rates, little of the rain is effective for crop growth.

By 1950 Egypt faced a water crisis driven by population increase. There was an urgent need to irrigate more land to feed the rapidly rising population. Urbanization and industrialization put further pressure on water demand and at the same time there was a need to develop more power. Clearly the old Aswan Dam could no longer provide sufficient water, and its height had already been increased twice.

The only way forward to increase the water supply was to build a new higher dam impounding a much larger reservoir. In 1954 it was agreed that the High Aswan Dam project was technically and economically feasible. An agreement between Egypt and Sudan over the amount of Nile water allocated to each country was reached in 1959. Work started on the project in 1960 and by 1968 the reservoir, Lake Nasser, was full.

The High Aswan Dam Scheme

Key Facts

The most suitable site for a dam:

- the valley is narrow and deep from Aswan upstream to Wadi Halfa where the Nile enters Egypt
- Lake Nasser is 500 km long, averages 10 km in width, has a maximum depth of 98 m and can store 164 million cubic metres of water, that is three times Egypt's annual share of the Nile flow
- this section of the valley was of little agricultural value and there were no major settlements
- Aswan granite could be used in the construction of the dam

Figure 4 Key facts about the High Aswan Dam

Over to you

1. The Nile is an international river (flowing through a number of countries) and Egypt is the country furthest downstream. Identify the countries in its drainage basin (**Figure 3**). Do you feel that even with the new reservoir Egypt has a secure supply of water?
2. Why could drought in Ethiopia seriously affect Egypt's water supply?
3. Egypt has vast water resources deep underground. Suggest why these have not been developed.

2.5 The High Aswan Dam – do the benefits outweigh the disadvantages?

What are the benefits?

The High Aswan Dam is a multipurpose scheme. Its benefits are not just local. Virtually all the people (60 million) have benefited, and the Egyptian economy has grown fast. Its benefits are ongoing. The total control of the Nile is the key factor.

A reliable source of water has brought untold benefits. Without this it would not have been possible to improve the quality of life for the people (social benefits) and neither would the country have been able to make such good economic progress. **Figure 1** shows the social and economic benefits the dam has brought.

Figure 1 Social and economic benefits of the High Aswan Dam

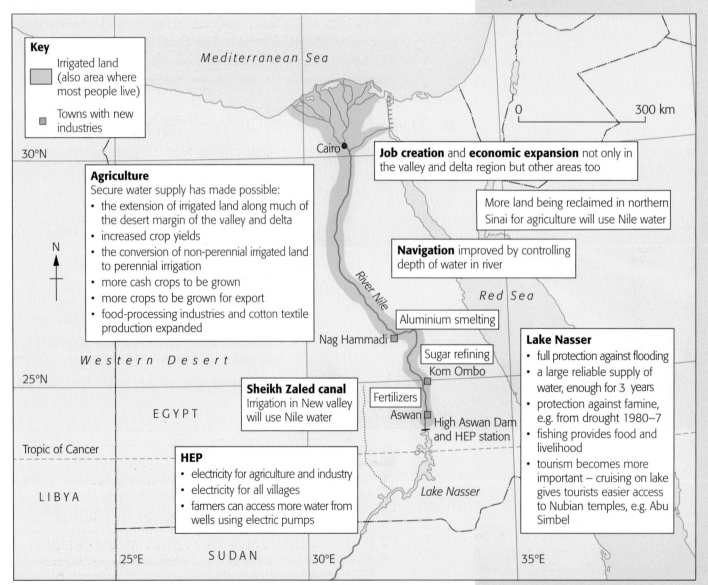

Study 2 Demand is rising. What can we do?

What are the disadvantages?

Building any major dam which benefits a country's economy can cause social problems for people living in the reservoir area and downstream. It also can have a negative impact on the environment. But note the High Aswan Dam was built before people were so concerned about the environment. And even now that we are more environmentally aware, for most LEDCs developing their economy has greater priority than protecting their environment.

Figure 2 Environmental and social costs of the High Aswan Dam

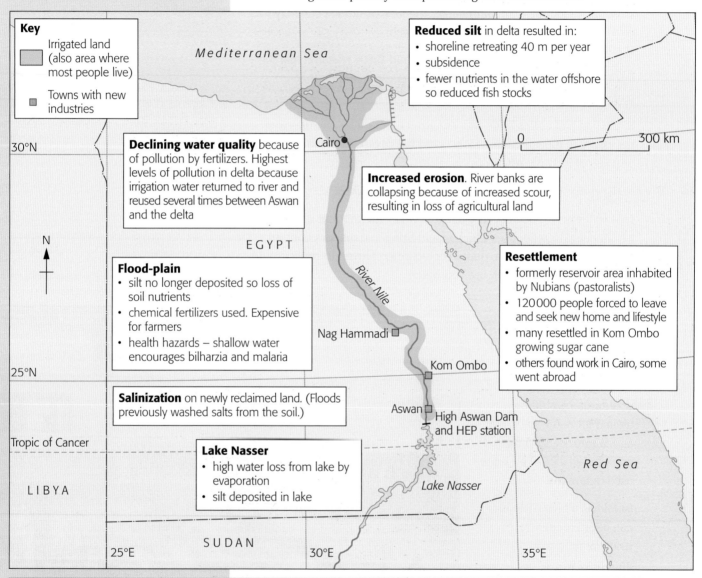

Over to you

1 Draw a diagram with water at the centre and show how the benefits brought by the High Aswan Dam are all water-related. Also show the links between the benefits. Highlight the social benefits.

2 Explain how the scheme has brought benefit to the whole country and not just the Nile valley and delta. Name an area which has developed in the last 30 years.

3 Why do LEDCs need to consider the social and economic advantages of such schemes first, rather than the environmental losses?

4 In water-short countries difficult decisions will have to be made in future. Should they build major water projects providing water, hydro power and flood control, enabling greater economic growth when large numbers of people have to be resettled and the environment suffers? What do you think?

2.6 Can we use our water more efficiently?

Water supplies can be increased by using water more efficiently, so cutting back on demand. This approach is known as **demand management**.

■ The problem for Israel

Israel is a water-scarce country. It is dependent on rainfall which is seasonal, inadequate and unreliable. The lack of storage capacity to cope with droughts is another major problem.

A re-evaluation of water resources in the mid-1980s revealed that the resources were only half the amount earlier estimated. In the meantime, population had doubled.

In the first half of the 1990s Israel realized that radical changes were necessary to meet demand. How could the government make the best use of its water resources?

■ What were the government's plans?

- Less water would be allocated to agriculture. The amount saved could be allocated to other uses which would bring in a higher economic return. For example, more people could be employed in services and high-tech industries which use less water.

Figure 1 More efficient use of water with a high-tech automated irrigation system

- A domestic water tariff would be imposed to make householders use their water more sparingly.

- Domestic and industrial waste water would be treated for re-use in agriculture. By 1995 this contributed 25% of all water used in farming.

- Israel has a high GDP and can afford to import essential foodstuffs. By importing wheat for example, which requires a lot of water to produce, Israel is importing **virtual water**. Accessing water through trading like this is much cheaper than importing water by pipeline, tanker or water bags.

Study 2 Demand is rising. What can we do?

How could stakeholders help?

- Householders could reduce consumption by using water-saving appliances.
- Industry could help by installing water-efficient equipment
- Agriculture could make further savings through more efficient irrigation methods. Drip irrigation not only directs water at the plants' roots but can also deliver fertilizer at the same time.

The future?

Drought forces Government to seek new supplies
April 2000

AFTER TWO SUCCESSIVE DRY WINTERS Parliament has finally given the go-ahead for the building of a desalination plant to be built on the Mediterranean coast at Ashkelon. It will be sited next to the power station. The plant, which will desalinate 50 million m³ of seawater a year, will cost approximately $150 million to build.

But it will be three years before the plant will be operational. In the mean time the government has negotiated with water-rich Turkey to buy 45 million m³ of fresh water a year over the next five years.

Other government measures to deal with the ongoing crisis focus on the speeding up of projects to recycle waste water for agriculture and purifying well water. Allocation of water to agriculture has been cut by 40% and farmers are facing a crisis.

Israel is in desperate need of exceptionally heavy rains in the coming winter. Lake Tiberias, the country's only natural reservoir, has shrunk to its lowest level for 100 years.

The Manavgat River and reservoir in Manavgat, Turkey, is a possible source of water for Israel

Over to you

1 **a** How does the water company which supplies your water deal with your water demand?
    b How do you try to reduce your water consumption at home?
2 Two solutions to Israel's water shortage are being found outside the country. What are they?
3 Desalination plants use a lot of power.
    a Suggest why most desalination plants are found in the Middle East oil-producing countries.
    b Explain the location of Israel's first desalination plant.
4 Desalinated water costs $1 per cubic metre. Suggest why it is mostly used for drinking purposes only.

WATER

Study 3
Is our use of water sustainable?

Demand for water is increasing in both MEDCs and LEDCs. It is possible to manage the demand in various ways, as we have seen on pp. 156–167. But the way we use and abuse water is something that needs careful management in order to make sure that we have a **sustainable** resource for the future.

3.1 How do we spoil our water?

In a word, pollution. More and more untreated waste water and effluent from industry is discharged into rivers. Runoff from fields carries agrochemicals and slurry into rivers. The result is a poisonous cocktail of pollutants and contaminants in the water. On navigable rivers pollution from shipping is yet another problem. The River Danube is a polluted European river.

The River Danube – environmental disaster

On 30th January 2000 disaster struck. Cyanide leaked from a gold-mine reservoir at Baia Mare reaching the River Tisza just over a week later. Two weeks after that it had reached the Danube and there was concern over water supply in Belgrade. Fish stocks in the Tisza River were poisoned, affecting many livelihoods. Water from the Tisza is also used for irrigation, so affecting agriculture.

Figure 1 Fish killed by pollution in the River Danube

Figure 2 Economic activities causing pollution

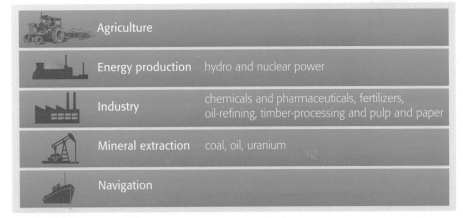

168

Study 3 Is our use of water sustainable?

■ Who is affected by pollution?

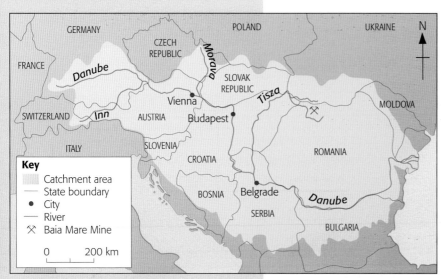

Figure 3 The Danube Basin

The Danube and its tributaries form Europe's second-largest river system. Within its basin live 85 million people who depend on the river for drinking water, livelihoods, tourism and leisure activities and navigation. It is not only quantity of water which is important. Clearly, quality is a key issue where drinking supplies and irrigation demands are involved.

But not only people are affected. The environment is degraded and half the delta is a World Heritage Site noted for its wildlife, especially pelicans.

■ Who initiates environmental action?

As an international river this has been a major problem. It has been compounded by the fact that the countries downstream of Austria are not so wealthy and do not have the financial resources to commit to environmental improvements. But, following the political changes which occurred in the early 1990s, the Danube countries agreed on a plan of action to clean up the river. The Danube River Protection Convention (agreement) came into force in October 1998.

Figure 4 Projects to improve the Danube

DANUBE
- Sustainable use of water resources
- Contaminants and human health
- Sustainable land use: reducing pollution from animal farms
- Wetlands and nature conservation
- Transnational monitoring network to provide information on pollution loads
- Accident emergency warning system
- Public awareness and participation – free newsletter 'Danube Watch'

■ What action is being taken?

A Strategic Action Plan was set up whose goals are to:
- reduce the negative impacts of activities
- maintain and improve water quality
- control hazards from accidental spills
- develop regional water management co-operation.

These goals would be achieved through various projects as shown in **Figure 4**.

Over to you

1. **a** Why is water pollution so difficult to deal with? Quote examples to support your answer.
   b What extra problems are there when a river flows through more than one country?
2. How will projects such as those in **Figure 4** help both people and the environment?
3. Use the web site www.environment-agency.gov.uk to find out about river pollution in the UK. Try to find answers to the following questions:
   a Which rivers are most at risk?
   b Which firms are most guilty of illegal pollution?
   c What type of pollution affects our rivers and what are its effects?
   d Which rivers have recently been cleaned up?
4. River quality is one of 15 indicators of quality of life in the UK. Write a short paragraph to explain why you think this is.

3.2 Water use – social and environmental disaster

What happened to the Aral Sea?

The Aral Sea in Central Asia is shrinking. **Figure 1** shows how the area of the sea has shrunk since 1960, and how the level of the sea has fallen.

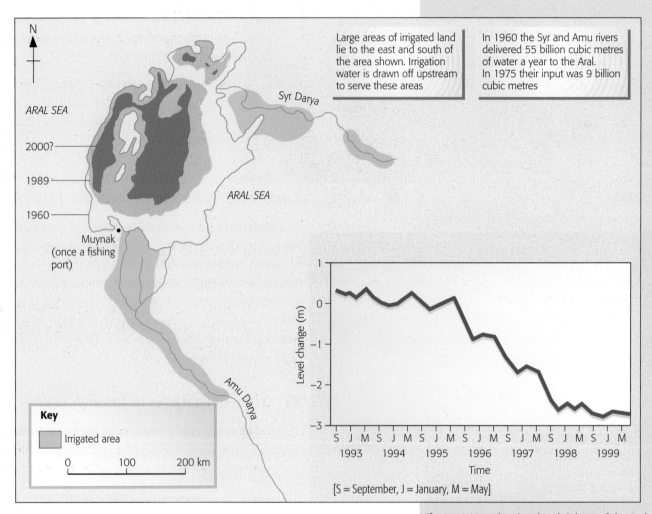

Figure 1 Map showing the shrinkage of the Aral Sea and graph showing the fall in its level

Why did it happen?

In 1950 the former Soviet Union decided to grow more cotton in the Aral basin, which lies in the Central Asian desert. This meant extending the area under cotton and the canal system to irrigate it. This involved the construction of a major canal, as well as lengthening others. Cotton is a water-intensive crop. It needs six applications of fertilizer (during cultivation) which is applied with water. But much water is also needed to spray it against pests and disease when it is the only crop grown (monoculture). Extra water was needed to grow more cotton, but much of it was wasted as it flowed along unlined canals.

Further high demands for water were made to grow rice to feed the population of the area, which had trebled as more people were brought in to cultivate the cotton.

Study 3 Is our use of water sustainable?

■ Why is it an environmental disaster?

Figure 2 A boat lies high and dry as a result of the sea shrinking

- The Syr and Amu rivers were almost drained of their water and so there was insufficient water flowing into the Aral Sea to counterbalance evaporation
- Water in the sea and aquifers have become increasingly saline
- Surface and groundwater supplies of water have become polluted
- Wetlands have been lost as flood plains of the rivers have dried out
- Salinization and waterlogging of soils has occurred because too much water was used in irrigation
- Wind-blown salt from the dried-out sea-bed is blown onto cropland and pasture
- Dust storms erode cropland
- The variety of wildlife has been reduced
- The fish have disappeared from the sea
- The climate has changed and now temperatures are more extreme. The number of frost-free days is less than 180, which is risky for growing cotton.

■ The future?

A plan, known as the Aral Vision, has been drawn up to restore sound ecological conditions in the area. This will not only regenerate the environment but will improve the quality of life for the inhabitants.

> **Over to you**
>
> **1 a** Explain why both the surface and groundwater supplies of water in the Aral basin have been polluted.
>   **b** What other action has depleted resources?
> **2** It has been a social disaster. Seventy five per cent of the population suffers from a chronic disease. What other problems do the people face?
> **3** It has been an environmental disaster. Draw a diagram to show how the effects are interlinked.
> **4** Saving the Aral Sea requires new water management. What would you suggest?

3.3 Global warning

Our increasing population, and use and abuse of water, are reducing the availability of a key resource and at the same time degrading the environment. These are global issues.

What is the damage?

In the **USA**: **Rivers** are under attack from over-abstraction and pollution.

- The Colorado no longer reaches the sea. With over 21 million people in the south-west USA dependent on it, there is no longer enough water to meet their demands.

- The Mississippi is badly polluted by industrial and agricultural runoff, and recreation craft and commercial barges stir up contaminants (PCBs) in the bottom sediments.

Wetlands have been drained.

- Water from the Everglades, once a shallow river, has been diverted to satisfy the demands of agriculture and residential development in south Florida. Polluted water and drained habitats have affected wildlife.

Aquifers are being mined and polluted.

- In the High Plains, the Olgallala aquifer is under threat as water is abstracted (taken) to irrigate wheat and thirsty crops such as maize and cotton. Centre pivot sprinklers and other powerful irrigation equipment remove water faster than rain can recharge groundwater stores.

In **China** in the Northern Plain.

- The demands for irrigation have lowered the water table.

- The Huang He river no longer reaches the sea the whole year round. Too much water has been abstracted upstream to meet the increasing industrial and residential demands.

- And now water is to be diverted from the Yangtze to meet the growing demands in the Beijing area.

Figure 1 Centre pivot irrigation system in Washington State, USA

Figure 2 Deforestation is responsible for the heavy silt load in the Upper Yangtze. Greater soil erosion has reduced water storage capacity and increased pollution

Study 3 Is our use of water sustainable?

In **Spain**:

- Donaña, Europe's most important wetland, is under threat from groundwater mining. Demands for water from outside the National Park for tourism, agriculture and residential development are the cause.

Climate change can also reduce supplies

In **Africa**:

Lake Chad is well known for its great fluctuations in level. The changes in level and surface area are determined mainly by rainfall in the Lake Chad catchment area. Human activity has had little impact. For much of the last century levels were generally low. But in 1962, after the wetter period in the 1950s, the lake surface reached its highest level of the century, 283 m above sea level, with a surface area of approximately 23 000 km^2. Then rainfall in the catchment area declined and by 1968 drought set in. By 1973 the lake surface had shrunk to 13 000 km^2 and by 1984 it was only 2000 km^2. The level in this year was so low that the intake pipes supplying water to an irrigation scheme begun in 1974 no longer reached the lake. The lake continued to recede until heavy rains fell in late 1999.

The future?

More big dams to satisfy our needs but degrade the environment?

Plans have been proposed and some implemented to make best use of resources to secure the needs of future generations, for example:

- Restoration plans involving river clean-up and the revival of the Everglades.
- Water management based on river catchments.
- Demand management.

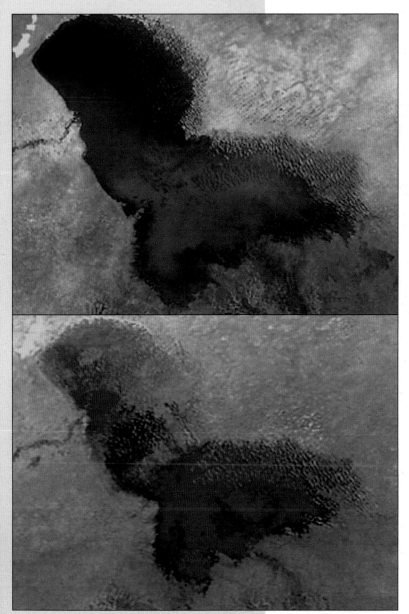

Figure 3 Disappearing Lake Chad shown from satellite images from October 1963 (top) and January 1997

Over to you

1 Can we avert a water crisis?
   It is an enormous challenge.
   We take water for granted. We just turn on a tap.
   In the future, will our glass be full or empty?

WEATHER AND CLIMATE

Study 1
How can weather and climate be a resource?

Weather is the short-term day-to-day state of the atmosphere. **Climate** is the average of weather conditions over a long period, at least 30 years.

In a country such as Great Britain, with some of the most changeable weather in the world, finding out what the weather is going to be like is a national hobby. Climates can be more predictable than day-to-day weather, but even our climate is changing.

1.1 Rain, sunshine, wind or snow?

Our interest in the weather is understandable. You need to know what it is going to be like when planning an outdoor event, or longer term when shops are planning sales. It's no good getting the winter coats on the racks in a store in the middle of a really hot spell in autumn!

Weather can be a resource, but short-term most people are much more worried about it being a hazard. Floods, droughts, gales, and blizzards increasingly cause misery to millions, as global population increases and more people are forced to live in more hazard-prone areas.

Climate is more long-term. The long dry hot summers of Mediterranean areas are predictable. But others, such as the monsoon in India, are only predictable up to a point. Exactly when, where, and how much rain will come in summer is of vital importance to Indian rice farmers.

Newspapers are full of doom and gloom as to how **global warming** is making nearly all climates more unpredictable. Of course, that's bad news if you want to use the climate as a resource – for farming, for tourism and leisure activities, and for energy use. Future management and control of this resource may become even more difficult. Can we really *control* climate, or do we just get very efficient at working with it? Some people would argue that capital and technology are the complete answer. We shall see.

The constantly high temperatures and abundant rainfall mean farmers can grow three crops a year.

Guaranteed winter sunshine and only 5 hrs flight away. The Gambia.

Lack of rain for two years in winter has left reservoirs dangerously low. Electricity cuts feared.

Plum prices skyrocket. Blossom destroyed by late spring frost.

Bumper harvest forecast. Spring rains and a sunny late summer did it.

We're thinking of purchasing solar panels but with sunshine records – will it be worth it?

The late winter storms have ruined the potato crop. We can't dig it up.

El Niño has messed everything up. The storms and floods ruined our peak tourism season.

Figure 1 Lush crops or flooded fields? It depends on the weather and climate

Shame about the weather isn't it?

THE PERCEIVED PROBLEM
To holidaymakers huddled in coats from Blackpool to Bournemouth there was just one question on their bluish lips, 'when can we go home?'

While parts of the Mediterranean sweltered in a heatwave, with record highs of 40°C 'Cool Britannia' shivered. It was the fourth bad summer for the guest houses and hotels.

BRITAIN SIZZLES IN THE SUN
The problem is it may be for only 8 days a year. The reason? Only rarely is the UK influenced by a dry tropical air mass from the Sahara Desert. Usually our air masses come from the wet Atlantic.

Cor – what a scorcher!

Study 1 How can weather and climate be a resource?

How important is weather and climate for tourism?

Looking at the pictures in **Figure 2**, you wonder how the UK can develop a tourist industry with such a climate. Yet the UK is sixth in the world for tourist arrivals (after France, USA, Spain, Italy and China). Tourism is one of the UK's most important industries earning over £25 billion a year. On the other hand, the British are fourth in the world league tables for the amount they spend overseas, as sun seekers or snowbirds. Twenty million people now take their *main* holiday abroad. In addition, nearly 5 million people take an extra short break in the UK, but tourism has had to adapt as tastes change. Traditional resorts, like Blackpool or regions, such as the Lake District, have always been at the whim of the weather. Now, new resorts like Oasis Lake District are weatherproof with high-grade, all-weather leisure facilities.

Britain may be a world leader for heritage, cultural, scenic, or business tourism, but our unpredictable weather and climate are a 'great big zero'. Visitors often say, 'Wonderful country, shame about the weather'.

Whilst climate is perhaps *the* most important factor in the development of tourist **hot-spots**, there are many other factors such as scenery, traditions, and good quality facilities. Of course, in future, global warming might change all of this…

Figure 2 Turncoat weather

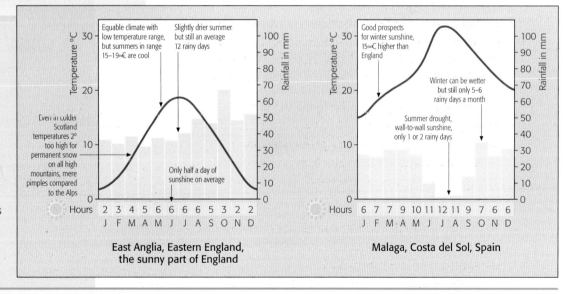

Figure 3 Climate graphs for UK and Spain. For sun-lust tourism there's no comparison

East Anglia, Eastern England, the sunny part of England

Malaga, Costa del Sol, Spain

Over to you

1 a Sort the statements in **Figure 1** into two groups – weather impacts, climate impacts.

 b Now sort the statements in another way – the costs and the benefits of weather and climate.

 c Try a third type of sorting – in which statements is there evidence of climate and weather being used as a resource?

2 Using the information on these two pages summarize why the UK is **a** good, and **b** bad, as a destination for tourists.

WEATHER AND CLIMATE

1.2 Attracting tourists – summer sun and winter fun

■ Summer sun

Climate is a major factor in **sun-lust tourism**, sometimes called '3S tourism' by the travel trade – sun, sand and sea. Growth in charter flights and marketing of package holidays has led to an enormous range of resorts for winter sun, e.g. the Gambia or Caribbean and for summer sun, e.g. Spain, Greece or Turkey in the Mediterranean. Some places like Dubai in the Arabian Desert, the 'hot' destination for the millennium, have 'all-year-round' sun.

In spite of skin cancer scares, many tourists look for holidays with guaranteed sunshine for 8–9 hours a day, generally with temperatures below 40°C. There you 'fry' and only need to move from air-conditioned hotel, to pool, to ice-cold drinks. Tourists will also put up with short sharp showers, but don't want rainy days. Warm sunny weather and high temperatures equal warm sea too!

Florida is an interesting example of sun-lust tourism – not for nothing is it named the 'Sunshine State'. Tourist areas try above all to avoid seasonality (i.e. only being open for part of the year) – think of all those wasted facilities if they are just used for the high season. Florida has all-year-round tourism as shown in **Figure 2** below.

Floridian tourism has suffered a few setbacks in recent years though, such as the murder of several foreign tourists in 1997. In 2000, a poor exchange rate from euros and sterling into dollars also increased costs for European tourists. But low-cost, long-haul flights and skilful marketing and development of attractions have made it a tourist mecca for families from Western Europe.

Even now, growth continues. The most recent boom is Caribbean cruising from Miami. 'Giant floating hotels' call every day and Miami is now the busiest passenger port in the world.

Figure 1 Cruising to the Caribbean in a 'giant floating hotel'

Figure 2 Florida – year-round tourism

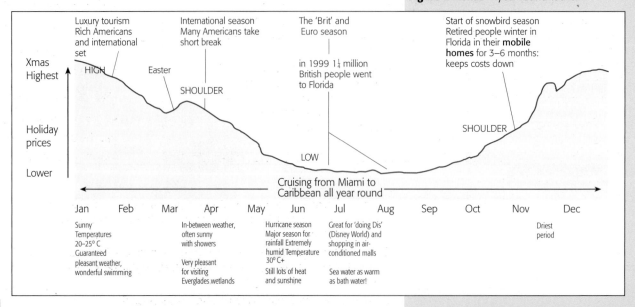

Study 1 How can weather and climate be a resource?

> Our problem is we all need different types of run. Our children need advanced runs, but I'm still at nursery slope stage and my wife has only been once before.

> We are weekend skiers so good access within 3–4 hours of home is vital. And you have to get back for work.

> The climate is very important for me. It was absolutely freezing with the chill factor in Norway. Also the lifts didn't work for three days because of the wind. Sunshine is such a bonus. Fog and mist are a disaster.

> For me, I go for high altitude resorts which guarantee snow straight out of the chalet. Here you can ski on the glaciers if all else fails.

> Quality of ski instructors – we need to be safe too. I'd never go anywhere with high avalanche risk. It's why we always go in January.

> What I like is that you buy a lift pass for the week – and it is a little less expensive here and you can get access to 200 km of snow. Such scenery and variety of runs.

> Although I enjoy skiing the atmosphere is vital for me. The boutiques, the hotels, the ice rinks, the après-ski.

GOOD SKI GUIDE
One of France's best purpose-built Ski Stations. 200 km of runs right from the doorstep are guaranteed to please. Ideal for beginners, intermediates and advanced skiers. Après-ski quiet.

Mountain Info

Beginners	excellent
Intermediates	excellent
Advanced	excellent
Snowboarding	good
Top station	3226 m
Pisted runs	200 km
Longest run	7 km
Direction of slopes	N, NE, NW
Max vertical drop	2026 m
Off-piste skiing	excellent
Cross country	15 km
Mountain restaurants	16
Glacier	Yes
Snow-making	Yes
Drags	29
Chairs	30
Gondolas	3
Cable cars	2
Funicular	1

Winter fun

The ski-holiday trade depends heavily on the right kind of environment for activities such as snowboarding and skiing. The Cairngorms ski resorts in Scotland have a very limited and unreliable snow season – there is usually little or no snow before Christmas, but when it does come blizzards are common, which block all access roads into the ski fields! Some resorts have such unpredictable climates that they rely on artificial snow to keep ski-runs open.

Snow cover is everything. What's important is:

- earliness of first snow,
- length of snow season,
- quality of snow cover,
- depth of snow cover,
- reliability of snow cover from year to year,
- size of snow-covered areas,
- shape of landscape so there can be ski-runs of all different grades of difficulty.

The high-altitude, modern purpose-built resort of Les Arcs nestles in the Isère Valley. It has always been a strong favourite among locals. The expansive skiing is set around the three satellite villages of Les Arcs 1600, 1800, and Arc 2000. It has a great snow record and with the Glacier du Varet at the peak of its ski area, snow is guaranteed. The wealth of slopes for all abilities is tremendous from the Aiguille Rouge at 3266 m to the tree-lined runs down to Villaroger.

The attraction of Les Arcs is the well-serviced ski area with some of the longest descents in the Alps, great value self-catering accommodation and pedestrianized streets.

Figure 3 Les Arcs in France: what a top ski resort has to offer

Over to you

Climate is a resource for tourism. For example, high-altitude sheltered sites can lead to the right kind of snow for winter sports, but a lot of other factors can be vital.

1. Look at **Figure 3** and the resort summary on it for Les Arcs in the French Alps. List the key features which skiers might see as very important when choosing a ski resort.
2. What do you think has made Les Arcs so successful?

1.3 Weather, climate and energy – rolling back for a new future

Some climatic elements such as wind, sun and precipitation (snow and rain) have enormous potential for energy generation. Today there is enormous interest in alternative energy sources – i.e. alternatives to fossil fuels such as coal and oil. These are considered 'green' (or environmentally friendly) and renewable. With rising concerns about air pollution, problems caused by **greenhouse gases** from the burning of fossil fuels, fears about the safety of nuclear power and so on, alternative energy sources seem an obvious way ahead. But are they? As you work through this section, you need to question this and whether it is right.

■ Alternative energy sources

- Are they commercially viable? In other words, can they produce large amounts of energy economically? If so, could we manage without fossil fuels or nuclear power?

- Are they a short-term prospect? Does current technology enable them to be used as a widespread alternative now or in the future? Many sources of power are variable, and so energy storage is a problem.

- Are they really 'green' or are there other environmental costs?

- What would be the best way to exploit and use these alternatives at present and in the future?

- Can they be used everywhere? The amount of solar energy reaching the earth exceeds the total of all other forms of energy. But its potential is far greater in some countries than others.

■ Hydro-electric power (HEP)

HEP is well established; in 2000 it provided 4% of the world's energy, and 20% of its electricity. Modern turbines are very efficient. In LEDCs HEP provides half of their total electricity demand. The technology is well understood and reliable, and is appropriate for most rivers at all scales, small or large. China has a policy of 'walking on two legs'. It has 200 000 mini hydro schemes that provide power for remote rural villages, as well as some of the world's largest mega dams, such as the Three Gorges Dam. Such large schemes are costly and have major impacts, such as flooding of settlements and quality farmland.

Climatic requirements for HEP schemes to be successful are for either steady rainfall, or annual snow or glacier melt to drive the turbines. But beware! New Zealand recently experienced short-term drought – a problem in a country that gets 90% of its electricity from HEP. So, there are advantages and disadvantages in using HEP.

Figure 1 Rolling back for a new future

The great dam scam
Has half a century of dam building done more harm than good?

MANY DAMS are failing to live up to expectations. Instead they make flooding worse, and cause havoc and conflict, says a report by the World Commission on Dams.

The report is the first to assess the world's dams – the biggest drain on aid budgets for the past 50 years, costing $4 billion a year in the 1980s. So far dam building has driven up to 80 million people from their homes.

Few dams have ever been looked at to see if benefits outweigh the costs.

A quarter of dams built to supply water deliver less than half the intended amount. In a tenth of old reservoirs, the build-up of silts has more than halved the storage capacity. What's more, by stopping the flow of silt downstream, dams reduce the fertility of flood plains and cause erosion of coastal deltas.

It's not all bad news, though. Dams irrigate fields that provide a sixth of world food production, while hydroelectric dams power many homes and factories. But the rural poor rarely benefit – only the urban and well-off people.

The report also concludes that some dams designed to prevent flooding actually make it worse. Such problems will worsen with climate change, it says.

The Dam Commission says dam construction is one of the major reasons for freshwater fish becoming extinct and bird species vanishing from flood plains.

Figure 2 The great dam scam

Study 1 How can weather and climate be a resource?

■ The answer is blowing in the wind - or is it?

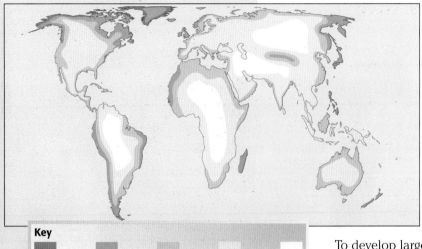

Key

Very high ← Potential → Very low

Figure 3 Worldwide distribution of wind energy potential

Wind power has long been used for driving simple machinery – windmills for grinding corn, or pumps for water, but for electricity the technology needed is more complex.

Winds blow over the earth's surface as a result of the uneven distribution of temperature and pressure. **Figure 3** shows how areas of high wind energy potential are widespread including many coastal and highland areas. However, while lots of areas experience strong wind regularly, it varies seasonally and from day to day, from calm days to those with destructive hurricane-force winds.

To develop large amounts of electricity from wind power requires huge numbers of wind turbines. The largest wind turbine in the world is offshore from Blyth, Northumberland. We would need over a thousand of these just to produce 1% of Britain's electricity!

The decision to develop wind power on a large scale depends not only on cost, but on also winning the environmental argument. Not everyone wants a wind turbine in their local area. Costs of design and building are high, together with maintenance, wear-and-tear. This makes energy production expensive, compared to the cost of producing electricity using coal, oil, gas, and nuclear fuels.

But the environmental arguments for wind power are strong, especially if countries are to meet their targets for cuts in carbon emissions by 2010. Against, there are largely local issues. 'A great idea, but where to put them? Not in my back yard!': this would be the 'NIMBY' view. The best sites for wind farms or parks are often hilltops or coasts in areas of beautiful scenery where they can be visually obtrusive. People argue the blades of the generators are a danger to birds, that they emit a slight whining noise, and that they can cause problems for TV reception. None of these concerns is proven, but each new proposal for a wind farm leads to a lot of local protest. Developments are occurring on land, and on a larger scale, such as the Mohave Desert in California. Increasingly, though, offshore sites are found to be better; Denmark hopes to get 50% of its power from wind by 2010.

Figure 4 Wind turbines in Yorkshire. What do *you* think of them?

Over to you

1. Draw a table with two columns – one headed 'HEP advantages', the other headed 'HEP disadvantages'. In the table identify as many advantages and disadvantages of HEP as you can. Consider its benefits, costs of development and of fuel, pollution and effects upon people and their life.
2. Using **Figure 2** and evidence you can find on the internet, summarize the arguments for and against building mega dams. Research two mega-dam projects to provide evidence of the costs and benefits of mega dams (use web page **www.dams.org**).
3. Draw a second table to show the advantages and disadvantages of wind power.
4. What are the arguments for and against producing 10% of UK electricity by wind by the year 2010? Research site **www.cat.org.uk** to help you.

1.4 How do weather and climate influence farming?

Using ecosystems

Farming involves using an **ecosystem** to provide food, but different types of farming use ecosystems in different ways. For example, traditionally, nomadic tribes have managed natural grassland ecosystems with their animals (**Figure 1**), but settled agriculture is very different. It alters natural ecosystems completely, by modifying plants, animals and the environment to produce food, as shown in **Figure 2**.

Figure 1 Nomadic people in Somalia

The effect of weather and climate

Some climates have great potential for food production, with high temperatures, and abundant rainfall. Away from the equator towards the poles, seasonality is a fact of farming life, and seasons, whether hot or cold, wet or dry, become more pronounced. Generally, these seasons are quite predictable, and farmers can prepare for them in advance; e.g. for the arrival of spring, or start of the rains. They can work with the climate to take advantage of seasons. Problems occur when farmers have to face severe unpredictable weather. No amount of capital or technology can overcome droughts or floods completely, but solutions are often very different depending on the level of a country's economic development. **Figure 3** shows how farmers react to problems of drought.

```
INPUTS
Water, sunlight, temperature,
seeds, stocks, plant foods
        ▼
      STORE
   Plant growth
   Animal growth
        ▼
 OUTPUT = FOOD
 Edible fruits, seeds,
 leaves, roots, milk, meat etc.
```

Figure 2 Using an ecosystem for food

In a less developed country	In a more developed country
Move to seek new land or look for a different type of work.	Use technology e.g. dry farming, diesel pumps.
Sell cattle immediately.	Consult farm improvement company to discuss schemes.
Use stored food in order to survive.	Modify landscape e.g. tree planting, terracing, contour ploughing.
Consult the village rainmaker and pray for rain.	Telephone insurance company to discuss claim for damage and crop failure.
Organize small-scale works e.g. ditches to bring water from river.	Have to talk to bank manager for further loans, or dig in to savings from previous years to buy up feedstuffs.
Change type of crop e.g. to hybrid millet or sorghum.	Sell up farm; make a new life in city.

Figure 3 How farmers react to drought

Study 1 How can weather and climate be a resource?

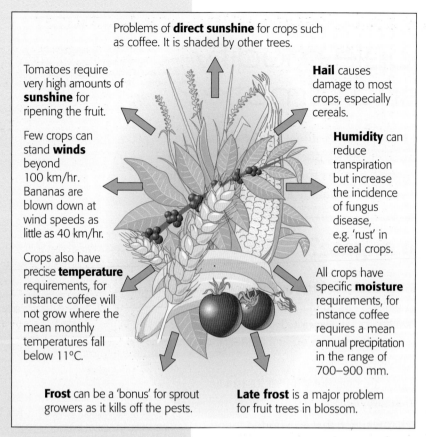

Problems of **direct sunshine** for crops such as coffee. It is shaded by other trees.

Tomatoes require very high amounts of **sunshine** for ripening the fruit.

Few crops can stand **winds** beyond 100 km/hr. Bananas are blown down at wind speeds as little as 40 km/hr.

Crops also have precise **temperature** requirements, for instance coffee will not grow where the mean monthly temperatures fall below 11°C.

Frost can be a 'bonus' for sprout growers as it kills off the pests.

Hail causes damage to most crops, especially cereals.

Humidity can reduce transpiration but increase the incidence of fungus disease, e.g. 'rust' in cereal crops.

All crops have specific **moisture** requirements, for instance coffee requires a mean annual precipitation in the range of 700–900 mm.

Late frost is a major problem for fruit trees in blossom.

Figure 4 Crop requirements

Temperature, moisture, sunlight, winds, evaporation rates, and frosts all affect crop growth, and can affect animals too, in terms of what they eat, for example grass-fed cattle. You need an average temperature of over 5°C for good grass growth. **Figure 4** shows how different aspects of weather and climate affect crops. Notice that what is good for some crops might be a real problem for others. For a crop to do well in an area, and make money for a farmer, it has to be well adapted to the environment. Some weather factors are consistent across a region, others such as frost or wind, or even sunlight can vary locally. Getting the best **microclimate** is absolutely vital for fruit growers. This is shown below in **Figure 5**.

Figure 5 A cross-section from south to north across a valley to summarize ideal conditions for orchard growth

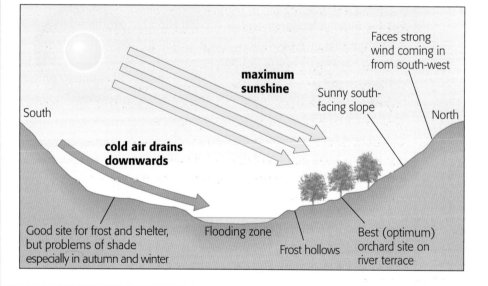

South — maximum sunshine — Faces strong wind coming in from south-west — Sunny south-facing slope — North

cold air drains downwards

Good site for frost and shelter, but problems of shade especially in autumn and winter — Flooding zone — Frost hollows — Best (optimum) orchard site on river terrace

Over to you

1. Look at **Figure 1** and describe the advantages and disadvantages for these people of nomadic farming, i.e. herding camels over a wide area, rather than on a settled farm.
2. How would nomadic farmers be working **a** with the ecosystem, **b** against the ecosystem?
3. What changes might take place that could threaten the ecosystem and their way of farming? Describe what these changes might mean.
4. Study **Figure 3** and suggest reasons why farmers react so differently to drought in LEDCs and MEDCs.
5. Using **Figure 4** and an atlas of the UK, identify ideal areas for farming. Refer to rainfall and temperature maps and identify areas which you think are ideal for **a** crop growing, **b** milk production, **c** fruit farming. Explain your answers.
6. On **Figure 5**, which is the best slope to grow crops – south-facing or north? Explain your answer.

181

WEATHER AND CLIMATE

Study 2
Should people modify weather and climate?

People can change weather and climate, either intentionally or unintentionally. Some attempts to alter the weather are deliberate, as we will see here. Others are unintentional, such as the development of **urban heat islands**, and **acid rain**, and can have dramatic and devastating effects.

2.1 Control at the farm gate?

Many people argue that the technology available to farmers allows them to overcome some weather problems and manage the limitations of climate. Look at the statements in **Figure 1**. They show some examples of how farmers could deliberately modify the effects of weather and climate in order to produce a crop.

- In **1** the costs of irrigating a desert area are high but worth it for Israel, which has limited resources of agricultural land.

- In **2** a simple change protects a high-value crop that is exported round the globe.

- In **3** the huge potential of genetic modification has produced drought-resistant wheat in Western USA, quick-ripening wheat for the shorter growing seasons of Northern Canada, wheat with short strong stalks to stand up to hail in Central USA, and wheat which is resistant to fungal disease. These are all working now; many more changes will be made to crops in future.

- Statement **4** however, makes another point. No-one grows bananas at the North Pole because it is not economically worthwhile. Iceland is a special case because it has almost free supplies of geothermal energy which is used to heat greenhouses in which bananas are grown. The cost of technology used to produce crops has to be balanced against the benefits of growing them.

Irrigating the desert

Several ways exist in which water can be brought to dry land that might otherwise be unproductive. This process is known as **irrigation**, several methods of which are shown in **Figure 2**.

Traditional methods of irrigation include:

- Basin irrigation, where the farmer works with the rainfall by allowing the river to flood fields ready for rice planting.

---

1 We have made the 'desert bloom'. Our bumper crops of citrus, celery and chrysanthemums go all round the world.
(Israel)

2 The giant belts of cypresses have been a stunning success. The kiwi fruit harvest is a bumper one and the harvest is safe – even from those very cold southerly winds.
(New Zealand)

3 We can ripen the tomatoes even without the sun now. It's all about genetic breeding. We can even get square tomatoes to pack easily.
(California)

4 We could even grow bananas at the North Pole – they already grow them in Iceland.
(USA News)

Figure 1 How farmers around the world are beating the climate

Figure 2 Irrigation methods

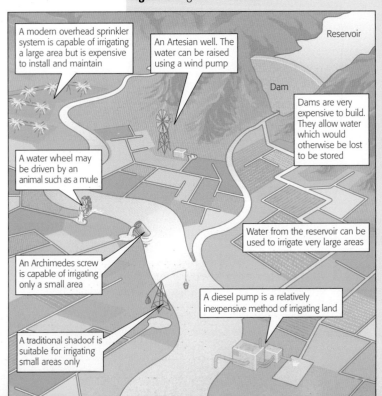

182

Study 2 Should people modify weather and climate?

In many areas of West Africa, on higher points of the flood plain which are only flooded for a short period, sorghum and millet are intercropped with a range of vegetables, or alternatively a second crop is planted in the dry season. Again, the farmer is really working with the rainfall.

Tanks and lifting devices such as a shadoof.

Other methods involve greater use of technology, for example tube wells, dams with irrigation channels, and sprinkler systems which help farmers to control water supply. However, these come at a cost economically, socially and environmentally. Page 178 looked at megadams, which have even greater costs and which bring problems. One common problem is salt concentrations in irrigation water where the climate is hot. Saline (or salty) soils rapidly reduce the ability of soils to grow crops.

■ Keeping the heat in

With some crops it is important that they don't get too cold. Farmers need to alter the effect of climate to ensure crops are warm enough.

Figure 3 shows a number of possible ways this is done. They either conserve the heat, by reducing the heat lost at night, or they add heat to warm the lowest layers of air. That's where the frost always is because cold dense air sinks.

Although some methods of keeping the heat in are successful – such as using blue polythene bags over hands of bananas to insulate them – many are expensive. You need about 100 heaters to cover just 1 ha of orchard. Think about the costs of running all those heaters for nights on end, even in the USA, the land of cheap oil. Of the methods shown, which do you think might be the most successful? Can this localized attempt to deliberately modify the **microclimate** of an orchard really work? In the short term and on a small scale, certainly, but don't forget that climate can be unpredictable, and unexpected frosts can kill orange blossom.

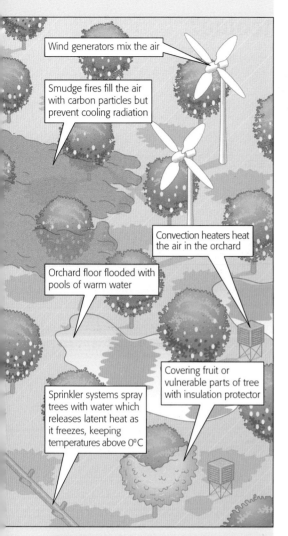

Figure 3 A citrus orchard showing methods of frost protection

Over to you

1 Make a copy of the table below and:
   a Describe how the weather can be modified at A, B, and C.
   b Research an example of each modification. Use the library, or the internet, and describe it in D to G. The first has been done to show you the need to be precise when using examples.
2 Consider all the methods of irrigation in **Figure 2** and compare them with dams which provide irrigation (see page 178).
   a Which methods are i) high and ii) low technology?
   b Which rely upon local materials and which ones have to be brought in from outside?
   c Which ones are best for LEDCs?
   Explain your answers.
3 Consider all the ways in which temperatures can be altered, using **Figure 3**. Which ones are practical, and for which places? Which other places, apart from farms, might want to purchase equipment to keep frost and fog away?
4 Discuss in pairs 'Should we modify weather and climate?' and write up your ideas in 200 words.

Weather element	How it can be modified	Example
Temperature	Use of greenhouses to raise temperatures and get the produce ready and ripe early.	Many crops such as tomatoes and flowers are grown in greenhouses in Jersey and Guernsey.
Sunshine	Shading bushes below trees can prevent too much sunshine.	D
Frost	A See **Figure 3**	E
Water (Rainfall)	B See **Figure 2**	F
Wind	C	G

WEATHER AND CLIMATE

2.2 Urban hotspots – an unintended outcome?

Cities alter the weather and have their own microclimates. The construction of every factory, motorway, office block and housing estate destroys one microclimate and creates another. Urbanization is increasing. In 1950 25% of the world's population lived in urban areas. In the year 2000 this was around 50% and by 2025 it could be 80%. More people will live in more and larger cities, especially in LEDCs.

Urban climates

People have a feeling that climates in these big cities are different. Their perceptions are shown below in **Figure 1**. What truth – if any – is there in these perceptions?

Figure 1 What do people say about climate in the city?

- Its just stifling in the summer. I dream of the cooler countryside. There's just no breeze here.
- My child gets asthma on cold winter days.
- There's hardly any snow in the city and when it does snow it causes the whole place to seize up.
- It's ever so hot even on a cold day. You can see all the people steaming!
- I'm sure we have more rain than they do – it could be all that soot.
- I'm convinced we have more thunderstorms. We even get them in the early morning, and we're getting more in winter too.
- I'm sure its very bad for the children's health. We have some very bad pollution. I cycle to work and I wear a smog mask.

Weather element	Comparing cities with rural areas
Annual mean °C	½–2 °C higher
Winter mean °C	1–3 °C higher
Solar radiation i.e. sunshine	15%–30% less
Precipitation amount	5–15% more
Thunderstorm frequency	16% more, especially in summer
Snow cover	4–5 days less
Fogs	60% more, especially in winter
Wind speeds, annual mean	25% lower
Calms	up to 20% more frequent
Cloudiness	10% more
Relative humidity	6% lower
Particulates	3 to 7 times greater
SO_2, CO_2 (pollutants)	2 to 200 times greater

Note The amount of difference depends on:
- size of city or the area covered by the city
- any industry present
- local factors such as relief, altitude
- closeness to water
- time of day
- season of year
- percentage of open urban space in city.

Figure 2 Average climate changes produced by cities

Study 2 Should people modify weather and climate?

Figure 3 The heat on the street: factors influencing urban heat islands

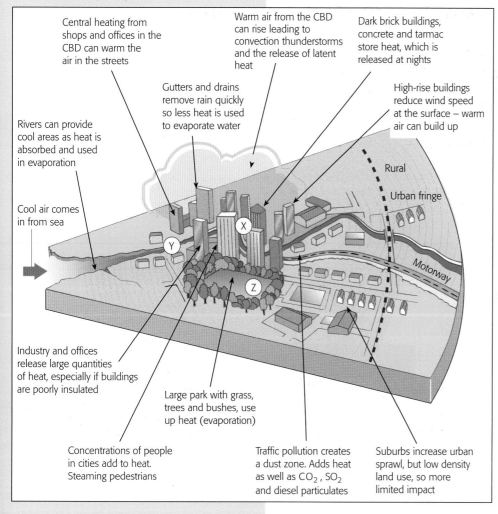

Cities – hot but wet

All cities have what is known as an Urban Heat Island, a zone of higher air temperatures found in and around urban areas. **Figure 3** identifies factors that help to explain this, while **Figure 4** shows how the more people there are, the hotter a city is!

Many cities also tend to be wetter than surrounding rural areas. Higher temperatures lead to rising air currents, which create unstable air. Dust from the city forms nuclei around which water vapour condenses and then falls as rain. The rough urban land surface traps depressions which can lead to heavy rain.

Figure 5 describes the effect of the urban heat island in America.

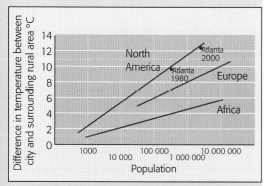

Figure 4 Population size, heat islands and the changing situation in Atlanta

Figure 5 Atlanta's own weather pattern

Growing American cities create their own weather

EXPLOSIVE urban growth is creating 'heat islands' so intense that they are establishing their own local weather systems.

The asphalt jungle and the lack of vegetation created by buildings makes cities many degrees hotter than they would otherwise be.

Huge heat 'domes' form over cities, triggering thunderstorms, increasing the production of polluting ozone, and raising local temperatures by an average of 5.5°C.

'Over Atlanta, the heat island is causing the city to create its own weather'. It's nicknamed Hot Lanta. 'At the end of July and the beginning of August, we have seen a series of thunder storms generated in the early hours of the morning – when no thunderstorm would normally occur – simply as a result of heat rising from the city.'

The reason for the effects are seen in satellite data which show the growth of the city. The images show that in the past 19 years Atlanta, one of the fastest-growing American cities, has lost 380 000 acres of tree cover, and gained 370 000 acres of single-family housing.

Replacement of the trees by roads and roofs means that heat is trapped during the day, and radiated back into space at night. The extra heat makes the city less habitable, forces air-conditioning units to work at full stretch, and increases the conversion of vehicle exhaust pollution into ozone.

Over to you

1 a Using **Figure 1** summarize what people feel about city climates.

b Do their views agree with results of research shown in **Figure 2**?

2 a Suggest reasons why area X is the hottest part of the heat island in **Figure 3**.

b Suggest reasons why areas Y and Z are both cooler than the surrounding areas.

3 Read **Figure 5** and explain how Atlanta 'creates its own weather'.

4 **Figure 4** shows how the heat island of Atlanta has developed since 1980. Summarize the changes since 1980, and explain why they have happened.

2.3 Acid rain – a spreading menace

What is acid rain?

We shouldn't really talk about acid rain. Instead we should say acid precipitation, as you can actually get acid snow and acid hail as well as acid rain! Burning fossil fuels, like coal, oil and natural gas releases sulphur dioxide (SO_2) and nitrogen oxides (NO_X), into the atmosphere. Gases escape into the upper atmosphere and react with water vapour in clouds to form weak sulphuric and nitric acid. In this way atmospheric pollution becomes incorporated into the water cycle, resulting in 'acid rain'. Hydrocarbons, which mostly come from car exhausts, also play a damaging part. High levels of low-level ozone in photochemical smog help to speed up chemical changes, which turn sulphates and nitrates into soluble acid.

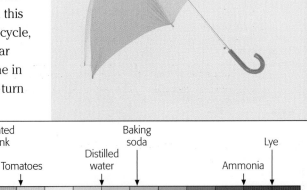

Figure 1 shows how acidity is measured on the pH scale. All precipitation is slightly acid, but acid precipitation is strongly acid. In the worst cases, it is as acidic as vinegar or lemon juice and can have a number of serious environmental effects. Note how the pH scale is logarithmic, so each point along the scale is ten times more acidic than the one before; pH4 is 10 times more acidic than pH5, 100 times more acid than pH6, and so on.

Figure 1 The pH scale

What effect does it have?

Dry deposition of chemicals usually takes place quite near the source, usually a factory or power station. People downwind from certain types of heavy industry complain of these effects – damaged washing, corroded garden gnomes, rusting bodywork or flaking paint on cars and all sorts of health problems.

Acid rain can be described as a spreading menace, because the winds can carry polluted water vapour in the clouds for many hundreds of kilometres around the world.

Figure 2 The acid rain cycle

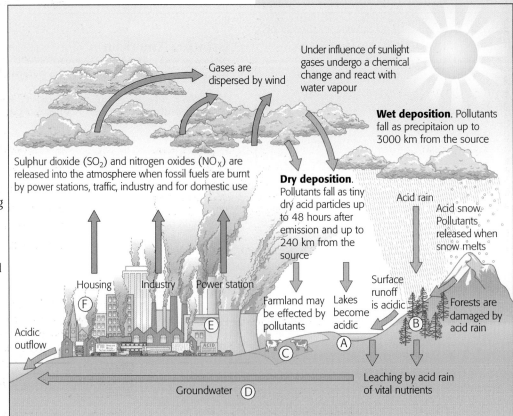

Study 2 Should people modify weather and climate?

Figure 3 Some of the serious effects of acid rain

	The effects of acid rain	Examples
A	**Lakes and rivers** In areas where the bedrocks weather slowly and do not contain calcium (lime), acid levels build up. Problems start at pH6 but when levels reach pH5 this kills fish – trout and salmon.	■ 20 000 acidified lakes in Sweden – 4000 are devoid of fish. ■ S. Norway – 80% of lakes are on a critical list. ■ Canada – 300 lakes in Ontario have a pH under 5 and are on the danger list.
B	**Forests** Acid rain washes out nutrients from the soil, killing many deciduous trees. It falls directly on leaves and bark, which damages even coniferous trees.	60% of Germany's forests e.g. the Black Forest, are affected by tree deaths. Coniferous forests in the Northern Appalachians (USA) are severely damaged.
C	**Farming** Crop yields are down as much as 20% in areas of acid rain. The respiratory system of farm animals is affected and leads to abortions in cattle.	In China huge areas of padi-rice have gone bright yellow in the Yangtze basin. Many deaths of young calves have been reported in Eastern Europe.
D	**Groundwater** As acid rain percolates downwards it concentrates in underground supplies.	20% of Swedish wells are now acidified and this causes a major problem for water supplies as the pipes are corroded.
E	**Buildings** Acid corrosion is eating away at metal, stone and wood. Many historic buildings are in danger, especially those built from limestone.	In India the Taj Mahal is threatened by acid pollution from Mumbai's industries. In Greece the Parthenon (temple) is threatened by pollution from Athens. In Poland acid corrosion is a danger to railway tracks.
F	**Health risks** Acid mists and smogs are common especially in industrial towns, particularly affecting young children and the elderly, with asthma, bronchitis etc.	In Northern Bohemia, industrial towns have a very poor health record as people literally choke on the fumes.

Figure 4 Solutions – planning for a cleaner future?

Remediation
- Spread lime in lakes
- Wash coal with lime
- Spray fields and forests with lime

Using energy wisely – conservation
- Energy conservation
- New renewable sources
- New fuel for cars e.g. hydrogen cells, electricity

Cutting vehicle emissions
- By reducing speed limits
- By cutting down on car use
- By cleaner fuels (low sulphur) and cleaner technology e.g. catalytic converters

Cutting SO_2 and NO_x emissions
- Putting scrubbers in chimneys
- Desulphurization of coal
- Moving away from fossil-fuel power stations

Monitoring levels of acid deposition
- Fining dirty factories
- Developing standards which are enforced by law

Planning a cleaner future
- Governments agree on production levels of SO_2 for the future. Decreased levels targeted of SO_2 and NO_x emissions

Highly acid rain has been falling over Europe for over a hundred years. As new areas in the world industrialize, there are more potential and widespread sources. Although attention is now focused on global warming, acid rain is nonetheless a current global problem. It is a particular concern in China and India, where rain with a pH of 2.5 (like pure lemon juice!) has been recorded.

Clearly, acid rain has some very serious effects both on the environment and people, shown in **Figure 3** above.

■ Can anything be done?

There are several strategies which can be used to do something about the acid rain problem. These are shown in **Figure 4**. Some solutions shown tackle the causes, others the effects. Some solutions can be carried out by individuals, some by companies, and yet others require government initiatives. Some solutions require international co-operation whereas others can be done locally. But most are expensive to put into effect. Acid rain is a difficult problem to solve, but one thing is certain – it won't go away.

Over to you

1 Using **Figure 3**:
   a mark on a world map the countries affected by acid rain
   b label the map with the effects of acid rain.
2 Look at the possible solutions shown in **Figure 4**:
   Draw up a table with 3 columns and 8 rows.
   In the first column add the heading 'Solutions' and then in each row write in each of the following features: 'which tackle causes,' 'which tackle effects,' 'can be carried out by individuals,' 'can be carried out by companies,' 'can be carried out by governments,' 'can be carried out by international co-operation,' 'can be carried out locally,' 'are expensive to put into effect.'
   In the second column add the heading 'examples from **Figure 4**'.
   In the third column add the heading 'likely or unlikely to happen? Why?' Now complete the table.

WEATHER AND CLIMATE

2.4 Acid rain in Europe

Figure 1 shows the death of a forest – the Black Forest (Schwarzwald) in Germany. It was considered one of the most beautiful areas in Germany but now it is a symbol of human abuse of the environment. The culprit was originally thought to be acid rain – but now there is more support for the idea of:

- acid soil which results from acid rain, or

- ozone pollution resulting from hydrocarbons from car exhausts which react in sunlight with nitrogen oxides to produce ozone.

Either way, the result is the same. In addition, a series of very hot dry summers also contributed to general tree sickness and may have accelerated the process of tree death.

The cause of the death of the Black Forest is the release of sulphur dioxide, nitrogen oxides, and hydrocarbons into the atmosphere over Europe. **Figure 2** shows the most heavily polluted areas in Europe at present – all former Communist smokestack industrial areas noted for their coal-based heavy industries. Acid rain is an old problem, not just from factories, but from coal- and oil-fired power stations, and a general use of fossil fuels for heating.

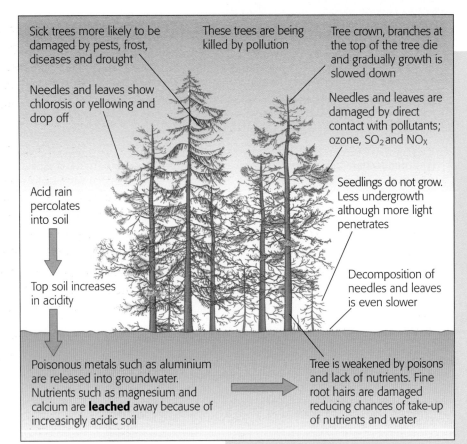

Figure 1 How human abuse of the environment kills off a forest

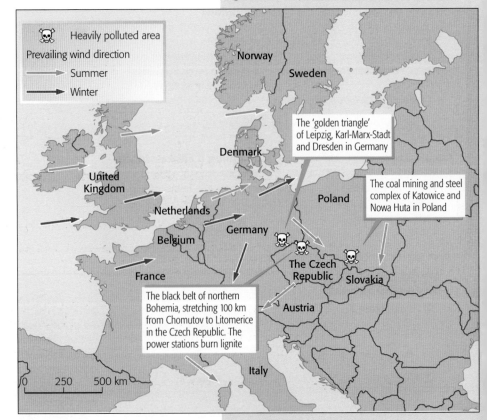

Figure 2 Heavily polluted areas of Europe in 2000

Study 2 Should people modify weather and climate?

Country	Sulphur deposited 1991 (100 tonnes as sulphur)	Percentage sulphur received from other countries
United Kingdom	6160	15
Netherlands	851	79
Denmark	612	75
Norway	1351	95
Sweden	1995	86
Belgium	872	53
France	5024	60
Germany	13 502	28
Poland	10 712	50
Switzerland	754	84
Italy	5371	40
Austria	1876	93
Former Czechoslovakia	6167	47

Figure 3 Sulphur depositions in Europe, 1991

Figure 4 Sulphur dioxide emissions in Europe, 1980–2000

Figure 3 shows how acid rain is a Europe-wide problem. Because of the impact of winds (see **Figure 2**), some countries such as Norway, received nearly all their sulphur dioxide from other countries – what a gift! Others, such as the UK, are luckier, as 85% of sulphur deposited actually came from within their own country and only 15% from outside. Why do you think this is?

What is Europe going to do?

Figure 4 shows some data for 1980, and each country's intentions about reducing their emissions at the time. You can see that there were varying levels of commitment. The solutions shown in **Figure 4** on page 187 show that a decision to reduce any sources could have major effects. Take one example: phase out coal and oil-fired (fossil fuel) power stations. Think what the effects would mean for the coal mining industry, or what a reduction in nuclear power stations might mean for areas such as Sellafield in rural Cumbria.

Another possible solution lies in abolishing the internal combustion engine, or altering people's motoring habits. You will have a chance to debate this further at the end of the chapter, as the causes of acid rain have many links to other pollution issues. Imagine the political risk for any government that tried to turn people away from their cars. You might remember the revolts all over the world about petrol prices in 2000. They could be small by comparison.

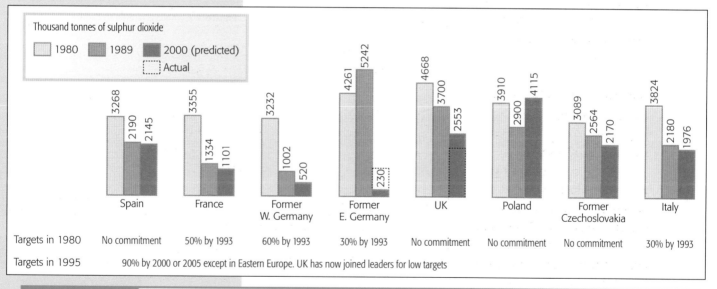

Over to you

1. Explain the distribution of the current 'hot spots' of heavily polluted areas in Europe. Imagine a map from the 1950s showing heavily-polluted areas. How different would it be?
2. Using **Figure 3** put the countries depositing sulphur in 1991 in rank order, making the one depositing the most sulphur number 1. Suggest reasons why the countries are in this order.
3. The UK has been termed the 'dirty man' of Europe for causing too much acid rain! Do you think Britain deserves the blame for much of Europe's acid rain? Weigh up the evidence from **Figures 2** to **4** and say whether you think this is fair.

Study 3
Global climate change

We've all heard about global warming – but is it really happening? What are its effects? And how will it affect us? Weather and climate is a resource which we can use or abuse. What we do has an impact now, and in the future. **Sustainability** may be our only option – acting today to conserve our planet for future generations.

3.1 Global warming – the planet's hottest problem

Global warming is a term used to describe the increase in global temperatures and is itself a 'hot' topic. The media are filled with dire predictions. A survey in 2000 showed that 20% of all environmental coverage in the press was about global warming. Acid rain is now 'yesterday's problem' – it was barely mentioned, the ozone hole, and saving the rainforest were second and third.

It is a heated controversy, as a number of questions have to be answered about global warming –

- Does it exist? If the answer is yes:

a Are we responsible for it, or can it be explained by natural causes?

b What exactly will its effects be?

At the moment just about every weather or climate hazard is blamed on global warming, but it is not responsible for *all* extreme weather events.

Climate change is nothing new

The earth is over 4500 million years old. In that time we have geological evidence of climate change with warmer periods, and very cold periods called 'Ice Ages'. In the last 1000 years, historical evidence shows climate change – such as a little 'ice age' in the UK in the seventeenth century. The Thames in London was completely frozen over and there were 'frost fairs'. **Figure 1** shows these climate changes in the last 1000 years; **Figure 2** shows how global temperatures have changed since 1860.

It is this most recent very obvious increase in documented global surface temperatures which leads on to another question – is there such a thing as global warming?

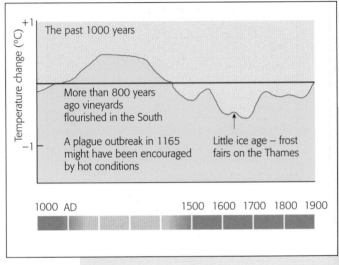

Figure 1 Climate changes in the UK in the last 1000 years

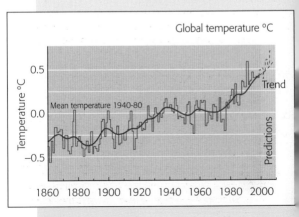

Figure 2 Global temperature changes since 1860

Study 3 Global climate change

Greenhouse Wars – is there such a thing as global warming?

Figure 3 The natural greenhouse effect and how it alters with increased carbon dioxide

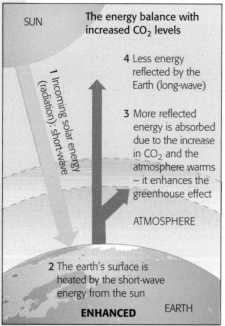

Some scientists do not believe in global warming. Whilst they accept that 'heating up' is happening at sea level, they say that satellite surveys of the upper atmosphere suggest that it may be cooling down. They say that the increase in surface temperatures could result from warming of the oceans. They also point out that a world warm-up should create more clouds, which would increase reflection of sunlight and therefore reduce temperatures.

Not surprisingly, views of the scientists who don't accept the idea of global warming are welcomed by some of the very powerful energy transnational companies (TNCs), as fossil fuels are the main culprit in raising levels of greenhouse gases.

The idea of natural causes of climatic change also has strong support. Is there any mileage in these ideas? For example, it may be a result of a change in the earth's orbit around the sun, part of a natural cycle of climate change, or due to changes caused by volcanic eruptions.

If, however, as a majority of scientists now think, human activities do have an effect on global warming, how exactly does this happen?

The greenhouse effect

Figure 3 shows the **greenhouse effect**, which is an entirely natural process. The ways in which heat is kept inside the atmosphere are natural. It is the **enhanced greenhouse effect**, brought about by a range of human activities, which is said to lead to global warming. Greenhouse gases trap heat, including CO_2, water vapour, CFCs, ozone, nitrogen oxides, and methane. All these gases have increased considerably over the last 250 years. Some are more important than others in their contribution to global warming.

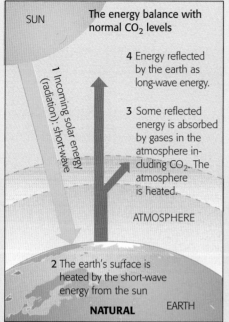

1 **Carbon dioxide** is the most important single factor in global warming. Estimates suggest it contributes nearly 60% of the greenhouse effect. It is produced by burning fossil fuels in power stations, in factories and in the home. It is also produced by road vehicles. A secondary source is from deforestation and the burning of tropical rainforests.

2 **Methane** is released from decaying organic matter such as peat bogs, swamps, rice fields, and waste dumps and above all farm animal dung. Methane contributes around 15% of the problem.

3 **CFCs** (chlorofluorocarbons) from aerosols, air conditioners, packaging, and refrigerators are the most damaging of the greenhouse gases. CFCs are about 20% of the problem.

4 **Nitrogen oxide** is emitted from fossil fuels, power stations, the use of agricultural fertilizers and from car exhausts. The amount of gases has grown sharply since the 1950s. This gas causes around 5% of the problem.

Figure 4 Some possible sources of greenhouse gases

> **Over to you:**
>
> 1 Using **Figure 2**, describe the climate trends shown in the last 140 years. What evidence does this provide for global warming?
>
> 2 a Summarize the arguments for and against the existence of global warming.
>
>   b Which argument seems more convincing to you?
>
> 3 Using the information in **Figure 4** suggest which features are likely to have the greatest impact on global warming in the future.

3.2 Global warming – should we worry about its effects?

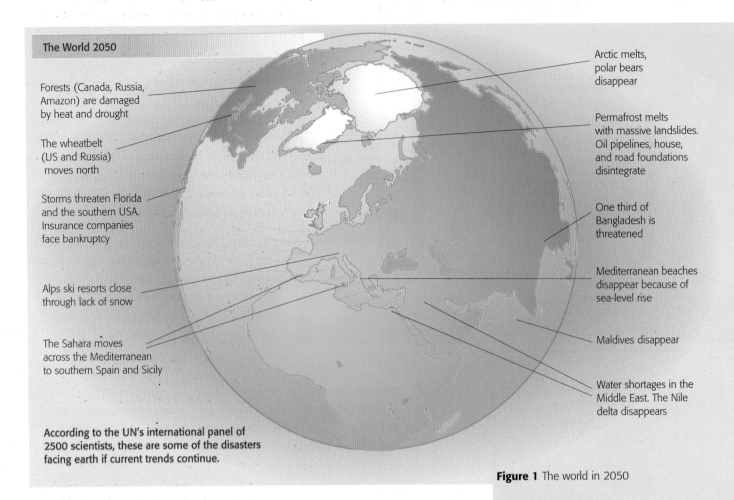

The World 2050

- Forests (Canada, Russia, Amazon) are damaged by heat and drought
- The wheatbelt (US and Russia) moves north
- Storms threaten Florida and the southern USA. Insurance companies face bankruptcy
- Alps ski resorts close through lack of snow
- The Sahara moves across the Mediterranean to southern Spain and Sicily
- Arctic melts, polar bears disappear
- Permafrost melts with massive landslides. Oil pipelines, house, and road foundations disintegrate
- One third of Bangladesh is threatened
- Mediterranean beaches disappear because of sea-level rise
- Maldives disappear
- Water shortages in the Middle East. The Nile delta disappears

According to the UN's international panel of 2500 scientists, these are some of the disasters facing earth if current trends continue.

Figure 1 The world in 2050

Figure 1 shows some of the possible disasters facing the earth if current trends in global warming continue. The predictions came from the UN's Panel on Climate Change – an international panel of 2500 scientists. As you can see there are some stark warnings. Would the Alpine ski resort on page 177 have to close because of lack of snow? However, even using the very latest computers, scientists still cannot agree on exactly by how many °C the temperature will rise and at what rate. Neither are they sure by how much ice sheets will melt, or the sea heat up, so estimates of rising sea levels (one of the most major global consequences) range from 80 metres (the extreme), to a more realistic 6 metres or even the latest estimate of 0.5 metres by 2100.

Rising sea levels are a major concern. A huge number of the world's peoples live in low-lying countries, often in very large cities such as Bangkok, or Dhaka, or even Mumbai – are all vulnerable to even a tiny rise in sea level. More storms could lead to more frequent and rougher seas, which could soon destroy the multi-million pound coastal defences, and cause coastal flooding to huge areas of countries such as the Netherlands.

Figure 2 Global warming – the main culprits

	CO_2 emissions	Population
	% of world's total, 1996	
US	25.0	4.7
Europe (inc. E Europe)	19.6	9.0
China	13.5	21.5
Former Soviet Union	10.2	5.0
Japan	5.6	2.2
India	3.6	16.3
UK	2.5	1.02
Republic of Korea	2.2	0.8
Canada	2.1	0.5
Australia	1.3	0.3

Study 3 Global climate change

Climate zones could move towards the poles by as much as 200 km, shifting entire ecosystems and agricultural zones with them. Here there could be winners and losers. Some northern countries could actually improve their harvests with milder winters, but others in southern Europe could become desertified with severe water shortage and crop failures.

But the big worry is that global warming is likely to hit hardest some of the poorest and most vulnerable countries in the world such as Bangladesh. Bangladesh has contributed virtually nothing to the atmosphere's supplies of greenhouse gases, yet it will be far more affected than the USA – the number one culprit (with 25% of the world's CO_2 emissions from under 5% of the world's people). It's a world thing – but an unjust one…

What will our climate be like?

The popular press have seized upon the idea of global warming with images of a sun-drenched future, with a Mediterranean way of life – as shown in **Figure 3**, but the reality is more uncertain.

The influential Climate Change Impact Group, based at the University of East Anglia state; 'As we see it at present in the next 50 years, it is likely that hot dry summers will increase and that chances of extremely cold winters would decrease, especially in SE England. But it will certainly be wetter in the North and West of Britain – especially in the winter.' So it's not all good …

Figure 3 An artists impression of how Britain could look in future

Over to you

1 Make a copy of the table below. Use **Figure 1** to complete the table.

Knock-on Effects of Global Warming	Example and Impact
Sea temperatures will rise – thermal expansion, causing a rise in sea level of 0.25–1.5 metres, which will have a huge effect on very low-lying areas.	
Land-based ice caps and glaciers, especially in polar areas will melt. This could raise sea levels up to 5 metres. It will devastate the ecology and economy of polar areas.	
The distribution of precipitation will alter – some places will be drier with a less reliable rainfall …	
… while others will become wetter and stormier.	

3.3 Global warming and the UK – are we winners or losers?

Figure 1 shows what are considered to be the most likely effects of global warming on the UK. But do we really know this will happen? There could be an alternative scenario. The 'Spitzbergen syndrome' (named after Spitzbergen out in the Arctic Ocean) could affect us, actually making the climate much cooler as the warm waters of the Gulf Stream are diverted away from the UK. Northern Britain could experience a tundra climate just like Spitzbergen. Instead of sunflowers in Sidcup, we could end up with igloos in Ilford.

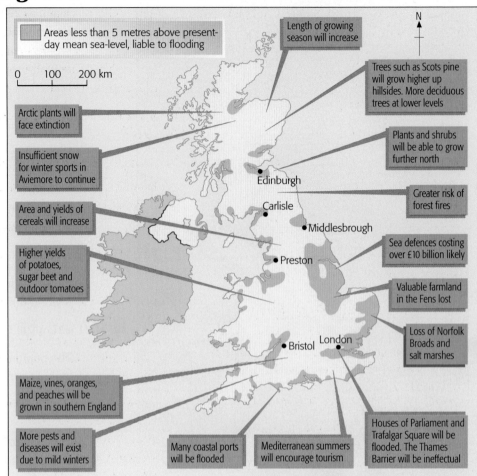

Figure 1 The effects of global warming on the UK

■ The time to act is now

If people are causing global warming, then in theory they should be able to do something about it. When looking into the future you have to consider different scenarios. One, business as usual, is sometimes called the 'status quo', and means 'leave things as they are'. The other, the 'sustainable option' means implementing the globally-agreed strategies designed to halt the production of greenhouse gases. This sustainable option might be very difficult to achieve because in the twenty-first century there will be massive economic development in China and India, which have over 40% of the world's population, meaning that they are likely to significantly increase their production of greenhouse gases.

Global warming is a world thing – so international co-operation is vital. At Kyoto in 1997, the rich countries agreed to cut greenhouse-gas levels by 5% below the levels of 1990 – to be achieved by 2008–2012. It doesn't sound a lot! They agreed to do it but then insisted that poorer countries followed suit. It was a compromise – and the follow-up conference at The Hague in 2000 was little better, even though it tried to improve on promises made in Kyoto.

Study 3 Global climate change

Country	Total emissions (million metric tonnes)	Emissions per person (tonnes)
USA	1392	5.35
China	848	0.76
Russia	462	3.22
Japan	303	2.53
Germany	237	2.96
India	225	0.31
United Kingdom	155	2.69
Ukraine	127	2.47
Canada	118	4.01
Italy	106	1.86

Figure 2 Who are the culprits?

1 The technological fix

Scientists can solve the potential problems of global warming. If CO_2 were pumped into the oceans, it could be taken up by plankton, eventually forming limestone, which is a long-term store for carbon.

2 Reducing carbon dioxide emissions

The only sure way to avoid the risk of global warming is to achieve a worldwide reduction in CO_2 levels, if all countries agree to policies that will reduce the use of fossil fuels. The world's wealthy countries will need to give aid to the poorer countries, and accept a lower standard of living.

3 Carry on as we are

Any reduction in the use of fossil fuels would lead to global recession and suffering in LEDCs. Not all countries would agree to cuts anyway. Theories that predict climate change are not proved, so why should we sacrifice our standards of living just because of a theory?

4 Actions work!

The UK used to have horrendous winter smogs in the days when coal was burned in homes, factories and on the railways. The Clean Air Acts of Parliament in the 1950s made everyone clean up and made city air cleaner. Taking action works!

5 Plant more trees

The solution is simple – plant more trees. Trees will store CO_2 and reduce levels in the atmosphere. Even if the trees are subsequently used for power generation, the next crop of trees will absorb the CO_2 released. Power companies which burn fossil fuels should be forced to contribute to forest projects and plant more trees.

Figure 3 Alternative strategies to deal with global warming

The environmental movements want to see reductions of over 80% in greenhouse gases. But for countries like the USA, a 5% reduction is very hard to deliver; since 1997 the American economy has grown hugely, using more and more oil. Study **Figure 2** – however you look at it, the USA is the main offender, and in 2001 reneged on the Kyoto agreement

How can nations reduce greenhouse gases?

- Change from fossil fuels to alternative energy sources.
- Invest in technology which uses less fuel e.g. cleaner car engines and fuel.
- Cut down the use of cars and develop public transport.
- Find alternatives! Electric cars and cars powered by hydrogen fuel cells are likely by 2025.
- Energy pricing – use taxes to price fuels so high that people cut down on usage.
- Control deforestation, and at the same time encourage tree planting to provide sinks to absorb and hold on to carbon from the atmosphere.

Politically, few countries could raise fuel taxes without losing votes, so The Hague 2000 seized on the idea of afforestation projects – both in the USA and in LEDCs – as an alternative. Scientists really are not sure whether this will work – it is no substitute for reducing emissions. The world agreed nothing new at The Hague 2000.

We may have to adapt to global warming, by building flood and coastal defences, irrigation and drainage systems, and new strains of crops to cope with new conditions. In the long term, populations may have to be kept away from areas vulnerable to flooding.

Over to you

1 Overall global warming in the UK will bring benefits, but the costs of it will be greater. In your view is Britain a winner or a loser? Use evidence from **Figure 1** to support your answer.

2 Decision-making exercise on global warming.
   a **Figure 3** shows five ways of dealing with global warming. Draw a table to show the advantages and disadvantages of all five.
   b Decide which would be your first choice. Justify your answer in 150 words.

To help you make your decision, use the Internet, and CD-ROMs from newspapers (*The Guardian*, *Economist* etc. November 2000 were very full of articles on global warming).

Useful Websites are:
   www.globalwarming.org   www.ipccc.ch
   www.newscientist.com/nsplus/insight/global/faq.html
   www.ncdc.noaa.ga/ol/climate/globalwarming

3.4 The crisis in the atmosphere – it's your decision

In this chapter you have considered how weather and climate can be a resource and how in a limited way people can manage and control this resource. Equally, you have seen how human actions can abuse the environment and lead to global problems, which are now so major that some type of global action must happen.

Some people believe that a sustainable future is the only option – that is, we have to act today so we conserve the planet for future generations. How popular is this idea?

Figure 1 Views about the environmental crisis in the atmosphere

Consumer

I'm worried about environmental problems, but I do whatever I can. I recycle cans and glass, I would never buy hardwood furniture and I buy CFC-free aerosols. But I don't think what I do really makes a difference. We live in the country so we have to run three cars – one is for work, my daughter has the second one for college and I use one to get around. We have a local bus, but I've never used it as it's so unreliable. I find the cost of motoring high as we use a lot of fuel, especially for the 4x4.

Haulage Contractor

I've had my own business for over 25 years and built it up to 10 lorries. The biggest change is the traffic jams. We take local farm products to a number of suppliers. I know that people blame lorries for a lot of noise and air pollution, but we can't do without them as people would soon complain if their shops were empty, and the petrol stations ran dry like in the fuel protest in September 2000. I think the only answers are to build more roads and allow bigger lorries on them, and give grants to hauliers like me to buy green-technology lorries. I'm not sure if I believe in global warming and anyway it's happening slowly. I'm 55 now so it won't affect me.

Young person

I don't care about the fuss that's made about the environment. We've learnt about it in school but I don't see how I can help. I know that global warming is supposed to have bad effects but it won't happen while I'm alive. I get a bit worked up when I see all the forests being cut down and all the pollution and that, but I don't think it's much to do with me – I never did anything to make that happen. No, I don't recycle cans, I just put them in the bin. I don't see what I can do really.

Green Person

I'm pleased that the government has at last seen sense and the environment is again a priority. Many green organizations such as Friends of the Earth have been monitoring the impact of human abuse on the environment for years. But we were ignored. The key to the problem is the car. All rich countries have too many cars and use them too often. The answer lies short term in providing quality public transport and long term in planning for sustainable cities, where people can walk to work and to shop. We also need more environmental education and developing alternative energy sources such as wind.

Senior Civil Servant

What we need is clear information. Many studies contain bits of good science but they are also the results of some imaginative thinking and a lot of guesswork. Although we accept that there is definitely an acid rain problem and probably the climate is going to warm up, we have to be very sure before we spend taxpayers' money. A lot of conservation schemes involve spending billions, possibly to save only millions. We are a democracy so people should be able to decide via the ballot box what they want governments to do. If you take energy – all we can do is to subsidize the (development) costs of alternative fuels such as wind. As you can see there is a lot of conflicting advice out there and our job is to manage it for the government.

Well Known Author

I haven't, I must confess, lost much sleep over the environment. We have more to worry about than changes in the climate. We should concentrate on things we might alter, such as the homeless, 10% of children below the poverty line, injustice in the world, rather than worry about unpredictable weather. My new year's resolution for the last three years is to put glass bottles and plastic bottles in separate bins for recycling – but I haven't actually done anything yet. Indeed I think I may even be in favour of global warming because they say that Southern England will become like Southern France – my favourite holiday place. My only worry is that at the age of 65 I won't be there when it is due to happen in 2050. I am worried about the ozone hole though – I've already had several minor skin cancer scares.

Study 3 Global climate change

■ Think and act local

Another very important idea is to think and act local. Round the world, local people can make individual or community decisions, which lead to action to reduce the abuse of the atmosphere. It may be a world thing, but to you the slogan of a well-known supermarket 'every little helps' is worth thinking about. You can join an environmental organization so that you can protest about local abuse. Or you can continue as you are, the choice is yours.

Figure 2 Think and act local: it can make a difference. (From top right) reforestation scheme in India; Buddhist monks in Thailand encircle a threatened forest; refrigerator dump in Sweden where CFC gases are removed and other parts recycled; protestors against motorway construction at Twyford Down, Hampshire; aluminium recycling, UK.

Over to you:

1 For each photo in **Figure 2** explain how local or individual actions could help prevent the abuse of the atmosphere.

2 Suggest one further local or individual action you might take to help the atmosphere.

3 **a** Read the six statements in **Figure 1**. Consider how much each person believes in sustainability by completing the table below. In the right-hand column, give each person a score out of 5 to show how much they believe in sustainability – '0' means they do not believe at all, '5' means they are totally committed to it.

Person	Evidence that they believe in sustainability	Score out of 5
1		
2		

b Write a seventh statement, that summarizes your own opinions about the environmental crisis in the atmosphere. Add your final evidence and score to the table above.

FARMING

Study 1
The impact of modern farming methods

Farming methods have changed significantly. This study looks at how and why they have changed, what farming is like today in the UK, and the effect that it has on the environment.

1.1 Foot and mouth – the last straw?

In February 2001, a vet reported an outbreak of foot and mouth disease in pigs on a farm at Heddon-on-the-Wall, Northumberland. Under UK law, all such cases have to be reported to the government, as foot and mouth disease is one of the most infectious diseases known in the farming industry. When diagnosed, certain animals (pigs, cattle, sheep, deer) have to be slaughtered in an attempt to contain the infection. The outbreak at Heddon-on-the-Wall was the first of many cases, and was the start of a period of several months in which many millions of livestock were destroyed.

Figure 1 shows the first week of the spread of foot and mouth disease in England and Wales. It already appeared to be widespread, and seemed to be caused by:

- several movements of pigs and cattle around the UK
- wind movements, which could carry the virus between neighbouring farms.

Figure 1 also shows how far the disease had spread within five weeks. Although at this stage the number of new cases was beginning to fall, it had by then exceeded the extent of the last major outbreak in 1967. Compare **Figures 1** and **2**. Notice that **Figure 2** shows total cases in 1967 whereas **Figure 1** is only the start of the 2001 outbreak. However, some things are clear – compared to 1967, the 2001 outbreak spread more quickly, was more widespread, infected more animals, and resulted in more animal deaths as a result of government policy to slaughter all animals close to infected farms.

Farming - an international business

The 2001 outbreak showed just how much farming is an international business now. As **Figure 3** shows, cattle and sheep are traded all over the country and into Europe. There are fewer, larger, abattoirs used in cattle slaughtering, so that cattle have to be taken further to be slaughtered for meat. The distances involved mean that farmers now transport livestock in larger lorries, managed and owned by cattle dealers. Dealers themselves play as big a role now as individual farmers in what we eat, and where it comes from.

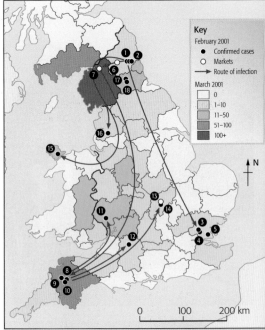

1. **Heddon-on-the-Wall Northumberland** (2–12 February) Foot and mouth virus infects pigs on Burnside Farm
2. **Ponteland, Northumberland** and **Westerhope, Tyne and Wear**. Animals on farms near Heddon-on-the-Wall are infected, probably as virus is carried on the wind
3. **Little Warley, Essex** (12–15 February) Cheale Meats abattoir receives infected pigs from Burnside Farm. Virus spreads to cattle on Old England Farm adjacent to Cheale Meats and owned by the same family
4. **Great Warley, Essex** Cattle on farm 5 km away from abattoir are infected
5. **Canewdon, Essex** Virus spreads to pigs at Greenacres farm near Southend
6. **Hexham** (13 February) Forty infected sheep from Northumberland are among 3500 sold at market
7. **Carlisle** (15 February) The infected sheep are sent to Longtown market, a holding centre
8. **Highampton, Devon** (16 February) The infected sheep arrive at Burdon Farm owned by exporter Willy Cleave
9. **Hatherleigh, Devon** (16–21 February) The infection spreads as sheep are moved to one of Mr Cleave's farms nearby
10. **Okehampton, Devon** Infection found on farm near Highampton
11. **Llancloudy, Herefordshire** Some infected sheep are sold to Hill Farm
12. **Bromham, Wiltshire** (21 February) Sheep from Highampton are taken to an abattoir where the disease is later confirmed
13. **Northampton** Another batch of infected sheep from Highampton are taken to market and sold to other UK dealers
14. **Wootton, Northamptonshire** Virus found on Blunts Farm
15. **Gaerwen, Anglesey** Abattoir receives infected animals from Yorkshire
16. **Withnell, Lancashire** Virus reaches Ollerton farm which received animals from Hexham market
17. **Wolsingham, Co Durham** Infection found here is traced to animals from Darlington market
18. **Wilton-le-Wear** Abattoir near Darlington infected

Figure 1 The numbers plot how foot and mouth disease travelled during the first weeks of the outbreak in February 2001. The shading shows how much the disease had spread five weeks later.

Study 1 The impact of modern farming methods

Pressures on farmers

For many farmers, the foot and mouth outbreak in 2001 was the final blow. Many farmers had left the industry during the 1990s as prices collapsed and difficulties in farming increased. Two factors caused this:

- **The spread of BSE**. In the mid-1980s, BSE spread among British cattle fed on sheep and rendered meat remains made by cattle feed companies. BSE was found to cross over into humans, leading to a fatal condition known as CJD. Although those affected by CJD are few so far, the scare was enough to cause a collapse in beef prices in the mid-1990s.

- **A general collapse in farm prices**. Most farm products are now sold at lower prices than in recent years, as shown in **Figure 4**. There are several reasons for this, but they include a rapid increase in the number of sheep kept, which has caused a sharp drop in price.

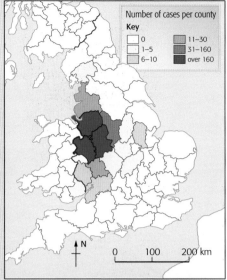

Figure 2 The foot and mouth disease outbreak of 1967: how was it different in 2001?

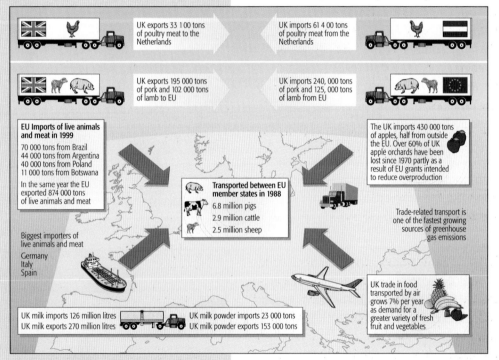

Figure 3 Meals on wheels – the mad dash as food moves around Europe and beyond

Figure 4 An auctioneer spells it out: what falling prices mean for farmers

FARMERS in wellington boots and green waxed jackets hung their arms over the bars taking notes. One or two looked more like City business men. The heifer was 19 months old and weighed 430 kg. The bidding was quick and decisive: the beast went for 79p a kilo.

'If you were here in the prime beef ring six or seven years ago,' the auctioneer said later, 'you would have seen farmers getting about 120p per kilo. That is why so many are going out of business. Four years ago, young female sheep would be going for £80-odd, and today they are averaging £30.'

Back in his office, the auctioneer took out his book covering the last few years. 'Last week, bulls were averaging 82.7p per kilo. Three years ago the average for bulls was 101.5p, and in October 1995, before BSE, the same bulls were fetching 134.2p. Think about that. There it is in black and white. You can't argue with those figures. Hellish.'

Over to you

1 a On an outline map of the UK, copy the 18 locations and arrows shown on **Figure 1**. Give your map a title.

b Use two different colours, to show which infections had probably arrived by road, and which by wind.

c Which seems to have been the more important way in which the disease spread, and why?

2 a Looking at **Figures 1** and **2** compare the extent of foot and mouth disease in 1967 with that of 2001. Mention:
- counties and areas most affected
- numbers of cases in areas most affected
- areas least affected.

b Suggest reasons why the outbreaks were so different.

c Suggest reasons why each outbreak spread mainly across the western side of the UK.

3 a Study **Figure 3** and suggest why links with Europe are now so great.

b What impact does this have on lorry size and on settlements along the routeways between the UK and Europe?

4 a Draw suitable graphs to illustrate the fall in farm prices shown in **Figure 4**.

b Explain the link between these prices and BSE.

5 In your own words, summarize the pressures on UK farmers in the early 21st century.

1.2 How are farming methods changing?

The small hamlet of Fairstead in Essex is shown on the map in **Figure 1**. It has its own church, and the focus of the hamlet used to be Fairstead Hall (the former manor house) and Fairstead Hall Farm. In 1967, the year of the previous foot-and-mouth outbreak, it was a lively dairy farm community.

- On the farm itself were two farmhouses, each inhabited by cowmen and their families who earned their living there.
- Elsewhere, another 10 houses were each lived in by other farm workers and their families.

In total, 12 houses with on average four people in each lived off one farm. Some people milked cows while others managed arable crops, such as cattle fodder for the winter (barley, potatoes etc.) or for sale (wheat and sugar beet).

In 1967, the last of the farm horses was still being used. Less than a mile away in each direction lay other farms with similar numbers of people. There were many jobs to do in the evenings and as over-time – women, children, and men alike would weed, and join in the harvest of sugar beet (**Figure 2**) or potatoes. There was always work for everyone.

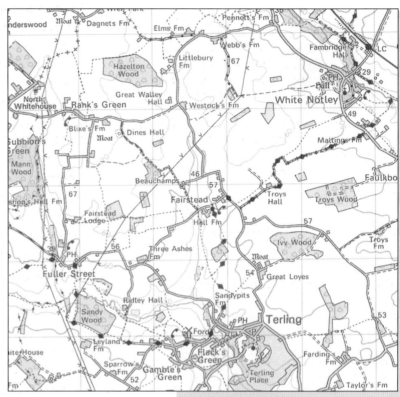

Figure 1 Fairstead in Essex

Now the entire farm and others around it are arable. One family now lives and works at Fairstead Hall Farm. Most people living in Fairstead now work elsewhere, and rarely in farming. Machinery has replaced people on the farm – tractors are more powerful, so can plough more quickly, and **mechanization** has taken away many of the jobs done by people. Fewer people are needed to farm the same area of land. All the farms in Fairstead have been merged into one unit now so that buying expensive machinery is worthwhile. There are environmental differences too. Hedges have been removed to enlarge the fields, so that less land is wasted. Life at Fairstead is very different (**Figure 3**).

Figure 2 Harvesting sugar beet in the 1950s: would you see this today?

Study 1 The impact of modern farming methods

■ Why are farms changing?

> *We see hardly anyone. The odd tractor driver comes down now and then, ploughs a field, and he's off. Several (of the old workers) got made redundant. One bloke was ploughing a field and they came up to him and said "Leave your tractor, you're redundant" and he had to go home there and then. They've taken all the birds away too – some of the hedges got grubbed up. It's like a dead hole here now.*
>
> Midge Baldry, talking about Fairstead Hall Farm in 2001

Figure 3 What changes in farming mean for local people

Figure 4 A success story: how change can be good news

- **EU membership**. In 1973, the UK joined the EU. Today the EU government in Brussels strongly influences UK farming. When large food surpluses occur, it uses grants to encourage farmers to change what they produce. For instance, milk **quotas** were used to stop over-production of milk. Farmers were given a strict limit of milk production and had to pay penalties if they exceeded it. To encourage change, the EU paid farmers grants to slaughter dairy cattle in return for ploughing up pasture for crops. Now, crops are over-produced, so the EU wants to reduce ploughed land; it pays '**set-aside**' grants for farmers not to cultivate land.

- **Animal welfare**. Many people are concerned about animal welfare, and the EU and UK governments have introduced laws to enforce minimum space and welfare requirements for animals. This makes meat production more expensive.

- **The role of supermarkets**. In 1998, the UK grocery market was worth £90 billion. Between them, the six largest supermarket chains (Asda, Morrisons, Safeway, Sainsbury's, Somerfield and Tesco) had 84% of this trade. They therefore control 84% of food sold in the UK, so they also control how it is produced and by whom, as **Figure 4** shows. For many farmers, selling at market is no longer a reality. They are tied to contracts with the main supermarket chains.

- **Farm ownership**. More farms are now '**agri-businesses**' – large farms owned by large companies or organizations with no connection with farming, such as insurance companies. Management pressures force reductions in costs in a drive for greater 'efficiency'. In reality, this means losing workers.

STAPLETON FARM is not far from Bideford, nearer Great Torrington, and there isn't a cow to be seen. No livestock, no fields, no manure, no tractors, just a small manufacturing unit that couldn't be doing better. This is the enterprise Sainsbury's put me on to when I asked about the partnerships with farming that mattered to them. This is the new thing.

I found Carol Duncan in a Portakabin she uses as a office. She was surrounded by Sainsbury's invoices and office stationary. Like her husband Peter, she considers herself a modern rural producer. 'I was delighted when we managed to get rid of the very last cow off this farm,' she said. 'That's the thing about cows, you know, they just poo all the time. We started our new business with three churns. We made yoghurt and started selling it.'

The Duncans now have more than thirty people working at Stapleton – chopping, grating, mixing, packaging, labelling, loading. The old farm buildings where the yoghurts and the ice-cream are produced look typical enough among high hedges of north Devon, yet inside each shed are silver machines and refrigerated rooms that are miles away from the world of cows.

Peter tells the story of the Sainsbury's development manager coming down to see them in 1998 as if relating a great oral ballad about a local battle or a love affair. 'She came down. I thought she seemed so fierce. They had already taken samples of our yoghurt away. They said they liked them. But when the woman came that day she just said : "I suppose you'd like to see these" and it was the artwork for the pots. They'd already decided we were going into business. I nearly fell off my chair.'

Carol laughed. "Yeah, and they said, how many of these could we produce a week?" They started aiming for 10 000 pots a week in a hundred stores. 'It was incredibly hard work,' Peter said. 'We were putting yoghurt into pots by hand and pressing lids on. They wanted more. We had to get better machinery. So it was off to the bank for £80,000. Come February 1999, we were doing 50 000–60 000 pots a week.'

Over to you

1 Make a large copy of the following table. On it:

   a list the pressures that farmers have faced, using examples on pages 198–199

   b show as many ways as you can in which they have changed as a result.

Pressures upon farmers	How they have changed

2 In what ways are the changes at Fairstead typical of what has happened in the rest of the UK? Give examples.

3 There are several controversial issues about farming. Consider each one of the following:

   a improving animal welfare on farms

   b supermarkets gaining greater control of food sales

   c the EU playing a bigger part in what farmers produce.

   For each one:

   i) describe why it is controversial

   ii) describe the arguments against it, and those in favour

   iii) give your own opinion.

FARMING

1.3 Farming as an industry

90% of British people live in towns or cities, and know little about farming. Whatever TV series such as 'Peak Practice' or 'Heartbeat' show about rural lifestyles, they do not show farming as an industry. Farming is now an 'agri-business' geared for profit, and needs a huge amount of investment.

Lynford House Farm

Lynford House Farm covers about 570 hectares in the Cambridgeshire Fens, north-west of Ely. Entirely arable, it is owned by a family partnership, Sears Bros. Ltd. The farm lies only just above sea level on flat fenland, with soils that are rich in mineral and organic content. With a low average annual rainfall of 559 mm, the area is ideal for arable farming.

The farm as a system

Each farm can be seen as a system. This means that whatever is put into the farm (inputs) will produce something (outputs). Economic outputs are the products of the farm, which Sears Bros. will sell. **Figure 3** shows a simple systems diagram, with inputs and outputs, and the farm itself as the centre of the system.

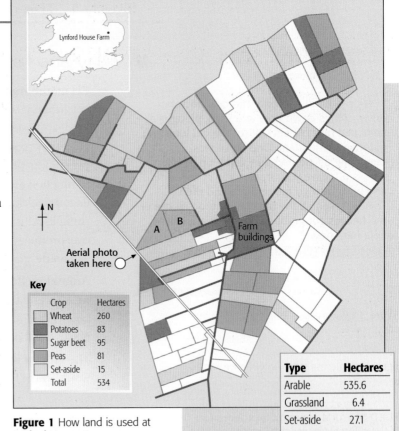

Figure 1 How land is used at Lynford House Farm

Key	
Crop	Hectares
Wheat	260
Potatoes	83
Sugar beet	95
Peas	81
Set-aside	15
Total	534

Type	Hectares
Arable	535.6
Grassland	6.4
Set-aside	27.1

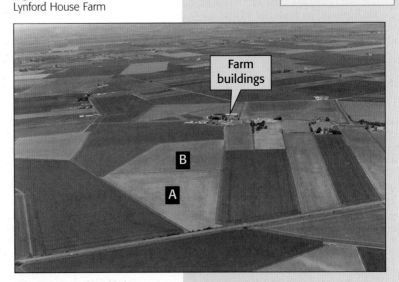

Figure 2 Part of Lynford House Farm from the air

Figure 3 A farming system

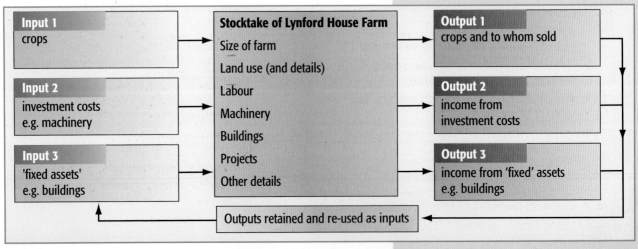

Study 1 The impact of modern farming methods

a Labour

The farm employs:
- 2 family members full-time.
- 3 full-time general farm workers.
- Contractors for muck spreading, lime spreading and vermin control.

b Machinery

The following major items were purchased over a ten-year period (1991–2000):

Item	Purchase cost	Year purchased
Combine harvester	£82 000	1991
Tractor 1	£33 250	2000
Tractor 2	£29 300	1993
Tractor 3	£25 000	1996
Potato harvester	£22 000	1995
4WD tractor	£15 913	1996
Roller	£12 000	1996
Trailers (2)	£10 000	1996
Harrow	£5 000	1999
Fertilizer spreader	£2 850	1999
Sugar beet harvester	£37 000	1999
Mower	£5 600	1999
Potato boxes	£8 200	1999
Total	£288 113 over ten years	

c Buildings

The farm has the following specialized buildings.
- Grain store with drying facility.
- Environmental cold stores for potatoes, plus other potato storage.
- General farm workshop.

d Purchase costs during the year

Item	Cost
Fertilizer: Solid nitrogen for grain and potato crops	£10 495
Liquid fertilizer for sugar beet	£4 790
Pesticides, fungicides, and herbicides	£43 538
Fuel (tractor diesel)	£7 671
	Total £66 494

The farm now belongs to a local farmers' buying group to gain purchasing power. In the first year it saved 20% on chemical costs.

Figure 4 The farm system at Lynford House Farm in facts and figures

e Products

In the period 1995–1999, the farm produced:

Crop	Yield in tonnes				Sold to...
	1999	1998	1996	1995	
Feed wheat	2252	1812	1497	1207	Grain co-operative
Seed wheat	0	150	465	465	Seed merchant, on contract
Sugar beet	4900	4500	4370	3453	British Sugar
Potatoes	0	3675	3000	2166	UK consumption and export to Ireland
Peas	0	0	400	400	Canning and freezing companies
Winter linseed	0	75			Contract for seed

f Subsidies

Set-aside payments and subsidies from the EU contribute approximately 6% of the farm's total income. These are grants paid to farmers who 'set aside' land previously used for wheat, and plant it with, for instance, trees and leave it out of cultivation. The payment is compensation for loss of income.

g Other enterprises

Income from traditional farming is becoming more and more unpredictable, but investment is needed to remain competitive and improve certainty about future income.

- 12 bungalows on the farm are let out, having been freed up as farm labour has been reduced. Rent now contributes 2% of farm income. Five of the bungalows have been sold off.
- The family is currently working with two companies, Eastern Generation and Wind Prospect, to build three wind generators on the farm, costing £1.5 million. They will produce enough electricity for more than 1000 homes. The farm will benefit from a steady income from this.

Over to you

1. **a** Describe the location of Lynford House Farm.
   b Using **Figures 1** and **2**, describe why this area is good for arable farming.
   c Draw an appropriate graph to illustrate the types of land use shown in **Figure 1**.

2. Make a large A4 copy of **Figure 3**. Complete the diagram, using **Figure 4**. Add any other boxes that you think are needed.

3. In what ways:
   a has the farming system changed in the past decade
   b might it change in the next ten years?

4. As a director of Sears Bros., would you argue for, or against, the proposals to:
   a build wind generators on the farm
   b sell off the remaining bungalows?
   Justify your ideas.

1.4 How does farming change the environment?

Lynford House Farm is like any other business: at the end of each year, it has to balance its books. As farm incomes fall in the UK, and the future becomes more difficult to predict, several changes are occurring at Lynford House Farm (**Figure 1**). To improve output, there are several possibilities, such as enlarging fields, mechanized hedge trimming, or the use of chemicals to improve growth or prevent hazards. But some of these have environmental impacts. As you read these two pages, consider ways in which the following impacts are linked.

■ Changing hedgerows

Hedges are important for two reasons:

a they are important **habitats** and corridors for insects, small mammals and reptiles that will not cross open fields.

b they form windbreaks between fields and provide shelter.

Between 1984 and 1993, 185 000 km of hedgerow in England and Wales were lost. To arable farmers, hedges are a nuisance and take up valuable land. Many would like to reduce them. Farmers have to notify their council if they wish to remove a hedge, because it could be 'important' for historical, ecological or archaeological reasons. But these reasons protect only 20% of hedgerows. This means that if farmers want to remove the other 80%, they can.

Poor management such as:

- ploughing very close to the hedge margin
- heavy pruning at the wrong time of year
- mechanical trimming using coarse-cut hedge cutters

is actually destroying more hedges than direct removal (**Figure 2**). Heavily trimmed hedges have fewer flowering plants and therefore less nectar for insects in the spring and fewer berries for birds in the autumn. This is one of the reasons for the loss of song thrushes (**Figure 3**) and sparrows, two of several species threatened by extinction. Sparrow numbers have fallen 60% since 1980.

- ♦ Reduce the number of farm products, concentrating upon wheat or sugar beet.
- ♦ Computerize farm operations e.g. crop recording, accounting, wages, and banking.
- ♦ Reduce the labour force.
- ♦ Rent out potato land and potato storage facilities to other growers. With 4300 tonnes of storage space available, Sears Bros. calculated that by renting out 80 hectares of land, and losing all the costs in irrigating and storing potatoes, they would earn £2000 per hectare.
- ♦ Harvest their own sugar beet, rather than employing a contractor.
- ♦ Instead of facing falling wheat prices each year, rent out land for vegetable growing to other farmers.
- ♦ The farm suffers from a lack of water. So install a 12 million gallon reservoir and a main distribution system around the farm.
- ♦ Improve farm security, joining the Cambridgeshire Countryside Watch scheme and invest further in improving security.

Figure 1 Ways of increasing income at Lynford House Farm

Figure 2 What happens when hedges are cut mechanically

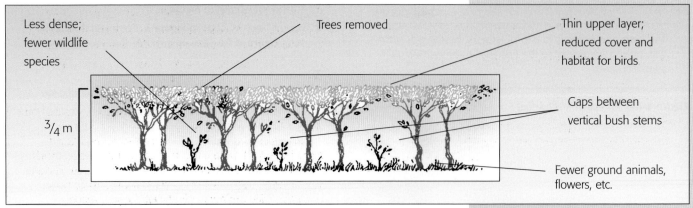

Study 1 The impact of modern farming methods

The effects of chemical sprays

Birds and animals are also affected by the use of herbicides (weed killers) and pesticides (insect killers). Not only does their use reduce the number and variety of species in fields and hedgerows – known as **biodiversity** – but it seriously reduces food supplies for those that rely upon 'weeds' or 'pests' as their food. It is not surprising that many British cities actually have a greater biodiversity than rural arable areas now.

Soil erosion

Repeated ploughing of soil, especially during autumn and winter months, exposes it to the weather at the wettest and windiest time of year. The Fens, in which Lynford House Farm is located, are flat with few features to break the force of any gales. The result is that **soil erosion** affects the Fens. Locals call such winds the 'Fen blow' and try to retrieve soils blown off the fields (**Figure 4**).

Figure 3 Species such as this song thrush are threatened when berries and insects disappear from the hedges

Figure 4 Exposed soil, few features – and suddenly fields take on a new look after a 'Fen blow'

Over to you

1. Consider all the options in **Figure 1** for increasing income at Lynford House Farm:
   a. on a limited budget, you have to decide upon three of these
   b. copy the table below and complete it

Choice	Risks	Good points

   c. identify the risks for each of the three choices
   d. identify the good points that could result if all goes well.

2. How would you convince your local council that more should be done to protect hedgerows in arable areas of the UK?
   Write a speech lasting 2 minutes (about 150 words) to be made at a local council meeting.

3. You are on work experience at your local office of the Environment Agency. The head of your section has asked you which of the three problems (hedgerows, sprays or soil erosion) described above is the most serious, and how it might be tackled. Write a report of approximately 350 words about the one you think is the most serious. Include diagrams in your report about tackling it. You may need to do some research to find out more information for your report.

FARMING

Study 2
Alternative methods of farming

What alternative methods of farming can be used? This study looks at two very different approaches – organic farming and the production of genetically modified food and crops. What is the future for these types of farming?

2.1 Organic farming – so what's the difference?

Aviaries Farm is located in Shepton Montague, near Wincanton in Somerset (**Figure 1**), and is a dairy and arable business. It is set in a rolling landscape with very little flat land, between 60 and 120 metres above sea level. Most soils are loam or clay and quite thin. Average annual rainfall is typical for this part of England, 863 mm per year. What makes this farm different is that it is organic. It does not use any artificial fertilizer, herbicide, or pesticide. It is an expanding business; the farm is unable to keep up with the demand for organic milk in the UK.

The farm as a system

Like Lynford House Farm, Aviaries Farm can be seen as a system. Like Lynford House Farm, Aviaries Farm is a business and sells whatever is produced. As you read through **Figure 4** consider which parts represent inputs, and which outputs.

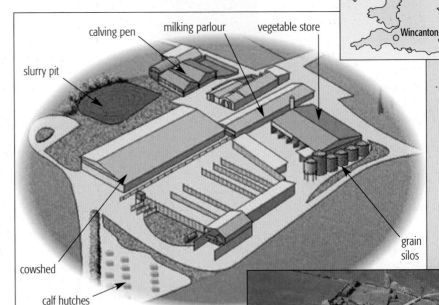

Figures 1 and **2** Aviaries Farm

Type	Hectares
Arable	205.5
Fodder crops	327.1
Set-aside	30.8
Woodland	21.5

Figure 3 Land use on Aviaries Farm

Over to you

1. Draw an appropriate graph to illustrate the types of land use in **Figure 3** and compare it with the graph you drew for Lynford House Farm on pages 202–203.

2. Make a large A4 copy of **Figure 3** on page 202 for Aviaries Farm. Complete the system diagram basing it upon the data given in **Figure 4** on page 207. Add any other boxes that you think are needed.

3. **a** Identify as many differences as you can in the different parts of the system between Lynford House Farm and Aviaries Farm. One way of doing this is to annotate another copy of the system diagram.
   b Suggest reasons for these differences.

4. In what ways:
   a has the farming system at Avaries Farm changed in the past ten years or so
   b might it change in the next ten years
   c is the future outlook for Aviaries Farm different from the outlook at Lynford House Farm? Why?

206

Study 2 Alternative methods of farming

a Land use in hectares

Type	Crop	2000	1999	1998	1997
Arable	Winter wheat	71.5	67.6	84.0	100.0
	Spring wheat	51.0	10.5	9.2	16.6
	Spring oats	0.0	28.5	0.0	0.0
	Triticale (a cross between rye and wheat)	83.0	85.0	16.4	8.9
	Potatoes	0.0	31.1	13.3	17.0
	Swedes (for human consumption)	0.0	6.8	9.0	10.9
	Sweetcorn	0.0	2.0	1.1	0.9
	Arable total	205.5	231.5	133.0	154.3
Fodder crops grown as seasonal fodder for the cows	Turnips	0.0	0.0	4.5	3.0
	Kale	0.0	0.0	0.0	12.1
	Fodder beet	12.1	18.2	21.0	12.1
	Grass ley	315.0	310.6	310.6	206.2
	Fodder crop total	327.1	328.8	336.1	233.4
Grass keep					28.0
Woodland, tracks etc.		21.5	21.5	21.5	21.0
Set-aside	Grass ley	12.5	24.7	8.0	8.6
	Woodland	2.1	0.0	0.0	2.1
	Other	16.2	0.0	0.0	0.4
	Set-aside total	30.8	24.7	8.0	11.1
Farm total		584.9	606.5	498.6	447.8

In 2000, there were 585 hectares, of which 266 hectares were owned and 319 hectares rented. Land area varies from year to year. The farm was increased by additional rentals in 1997 (28.5 hectares) and 1998 (160 hectares).

b Livestock

The farm has run the following Holstein Friesian livestock since 1996:

Type	2000	1996
Dairy cows	320	265
Young heifers (18–36 months old)	130	116
Replacements (less than 18 months old)	120	128
Bull	2	1
Total numbers	**572**	**510**

Each year about 50 older dairy cows are culled and 150 male calves are sold.

c Labour

The farm employs:
- 2 family members – one full-time, and one part-time.
- 8 full-time employees – a farm manager, 3 herdsmen, an arable foreman and 3 tractor drivers.
- A part-time secretary.
- Seasonal workers to weed and harvest.
- Contractors for ploughing and cultivation; hauling silage; baling straw; spreading manure; hedge trimming; ditching; and property maintenance.

d Machinery

The following major items have been purchased:
- Milking parlour (bought in 1995 for £100 000).
- 7 tractors (last one bought in 1998 for £25 000).
- Machinery – plough, rotary cultivator, seed drill, mower, grass rake, grass harvester, crop weeder.
- Combine harvester (bought second-hand in 1998 for £21 000), straw chopper and shredder.
- Mixer-feeder wagon.
- Materials handler (4 wheel drive, rough terrain).
- Dirty water irrigator.

e Buildings

The farm has the following specialized buildings:
- A large milking parlour, its associated yards, milk tanks, etc.
- 2 covered yards and cattle shed plus a general-purpose building.
- Cubicle housing and calf hutches.
- Silage pits.

Most were built in 1994–96 after the merger of two herds. £650 000 was invested in buildings, tracks and waste storage (for slurry, effluent and dung).

f Purchase costs

Fertilizer: the only fertiliser allowed is a small amount of slag, a long-term fertilizer, spread for phosphate or potash. Otherwise, farmyard manure is spread and dirty yard water sprinkled via the irrigator.

Pesticides: Supermarkets reject any deliveries containing aphids. Pest and weed control includes encouraging natural insect predators, crop rotation, and mechanical and hand weeding.

Stock feed: 240 tonnes of concentrate feed has been fed to the cows and young stock. In addition, 120 tonnes of homegrown feed was used and 12 tonnes of hay. They ate 4300 tonnes of silage, 650 tonnes of fodder beet, and 160 tonnes of straw.

Fuel: the farm uses 36 400 litres of agricultural diesel at 11p per litre.

g Products

The farm's main outputs are milk and wheat. In the period 1995–2000 the farm produced:

Annual yield	1999/00	1998/9	1997/8	1996/7	1995/6	Sold to...
Milk (litres)	2 000 000	1 850 000	1 770 000	1 600 000	1 201 432	Organic milk suppliers, Co-operative
Wheat (tonnes)	460	460	620	620	450	Grain merchant, miller

h Subsidies

Subsidies and grants contribute 5–7% of total income, including set-aside grants and some from conservation work, e.g. tree and hedge planting or wall maintenance.

i Other enterprises

Until the merger in 1994, there was a second dairy unit with the farmhouse, which has been converted to housing, creating six new dwellings.

Figure 4 The farm system at Aviaries Farm in facts and figures

FARMING

2.2 Organic farming – what are the issues?

What is the debate about organic farming? Since 1998, the number of organic farmers in the UK has doubled to about 2000. Britain produces only about 30% of the organic food consumed here but demand is rising sharply. On the other hand, a debate rages about how good organic food is.

Some opposing views come from the 'other' side – from people involved in agri-business, shown in **Figure 1**. But there are several other views, as these two pages show.

■ The economic motive

On one hand, the current farming crisis has had limited impact on organic farms. Crop and milk prices have remained steady and not fallen like those of conventional farms. This is not surprising. Supermarket customers pay 60 to 70% more for organic meat, vegetables and other foods. In February 2000, organic potatoes in one supermarket cost 285% more than non-organic kinds. Organic production costs are a third higher, but the average price difference on the shelves is double that.

■ Is organic food better for you?

There are debates about whether organic food tastes better or is better for you. Are people buying because organic food is better or because of their suspicions about how food is produced? As well as BSE scares in cattle, production methods in salmon farming are now a matter of debate. Growth of young salmon is accelerated by over-feeding, and antibiotics are used to fight infection.

'There's a role for organic farming but it is small. If agriculture were to go organic again, we would have an enormous food deficit. The products our company supplies increase crop yields by 35–40%. I think the organic movement is a western European luxury. The quality of food available now is the result of what the (chemical) industry has contributed to the protection of crops. One shouldn't forget the health hazards associated with organic food production, not to mention shovelling muck all over the vegetables.'

Michael Pragnell, CEO of Syngenta, a US agri-business supplying chemicals and GM products

Figure 1 Are organic foods really better?

Figure 2 Higher prices, more profits – organic is good for farmers

DR ANNA ROSS at the University of the West of England researched prices on a shopping basket that included fruit and vegetables, meat, poultry, and other groceries. Organic food cost 71% more at Tesco, 65% at Sainsbury's, 62% at Waitrose, and 60% at Somerfield. She believes that 'supermarkets exploit consumer demand with excessive price hikes on organic food.'

The Independent 6 February 2000

Figure 3 Sweden: farming organically… and successfully

IN SWEDEN, a salmonella epidemic that killed 200 people in 1953 prompted a review of agriculture. Since 1972, the country has aimed to have 'the cleanest agriculture in the world'. Pesticide use has been cut by 70%, pollution by fertilizers by 30%, organic farming has boomed, family farms have survived – and agriculture has prospered.

The Independent 25 March 2001

Study 2 Alternative methods of farming

Figure 4 Quality, safety, taste: why do people buy organic?

Figure 5 Organic farming and wildlife: too much of a good thing?

We pay for organic food because of a lack of trust

The government Food Standards Agency, declares that consumers are not getting value for money 'if they think they're buying food with extra nutritional quality or safety'. The difference in health risks between organic and conventional food is hardly measurable. In some cases like the risk of aflatoxins in organic peanuts, the benefits lie in the other direction. Nor is there evidence that organic food tastes better.

People pay more for organic food because they prefer to trust the Soil Association, the main body for organic produce, rather than the government's Food Standards Agency. Given the concern about beef and farmed salmon, this is understandable.

The Independent 2 September 2000

■ The effects on wildlife

Organic farming has positive effects on wildlife in terms of numbers and biodiversity. But there are two views of this, as **Figure 5** shows.

Organic farming 'is helping threatened wildlife'

ORGANIC FARMING is the saviour of Britain's rare wildlife, the Government's top conservation adviser said yesterday. Birds such as skylarks, song thrushes and swallows, flowers such as buttercups, and many smaller organisms such as insects, spiders, and earthworms do 'overwhelmingly better' in an organic environment.

Organic and conventional farms were compared for biodiversity across Britain. Tests in the Chilterns showed there were greater numbers of 18 species of flowers on organic farms. The buttercup – the most rapidly declining wildflower in Britain – was only present in organic fields. Earthworms, birds and spiders were also more abundant in an organic environment. In a survey of 92 species, 30 were found to be more common on organic farms compared with only four on conventional farms.

The Independent 6 January 2001

Aviaries Farm in Somerset (an organic farm) say they have:

- ... too many rabbits, grazing valuable grass. We have about 1000 rabbits eating 3.5 hectares of grass per year, which would keep 70 sheep.
- ... too many badgers. They eat slugs (good!) and worms (bad!), dung in crops, scrape at soil and make holes. Their numbers have grown excessively, with several hundred on the farm.
- ... birds of prey such as buzzards, hawks. Now protected, they have multiplied and kill more songbirds and pheasant chicks.

Over to you

1 Consider all the arguments on this page about organic farming. Make a copy of the table below and summarize the benefits and problems identified by both sides.

Issue	The arguments in favour of organic farming	The arguments against organic farming
The view of agri-business		
The economic motive		
Is organic food better for you?		
The effects on wildlife		
Other ideas that I have		

2 Which arguments seem strongest? Which ones are weakest? Why?

3 What are your opinions? Should organic farming be extended in future or just left to feed a minority market?

FARMING

2.3 Genetically modified foods – the case for ...

In 2000, the world's population passed 6 billion, having doubled in just over 40 years. The world's farmers have to produce more food now than they had to in the past, and it is clear that some parts of the world do not have the capacity to do this at present. Now imagine a solution to poverty and hunger, and a world in which the prospect of more people to feed is no longer a major challenge. That is the promise offered by the introduction of **genetically modified** (GM) food products – or at least by the companies who make them.

GM foods are being hailed as the great solution. They are claimed to be safe and environmentally friendly, reducing the need for chemicals in agriculture whilst still helping to feed the world's hungry. What truth is there in this claim? These next four pages will try to untangle myths and rumours about the GM debate.

What are we talking about?

Figure 1 shows an advert from Monsanto, a large American biotechnology company that advertises itself as providing 'agricultural solutions' – i.e. it solves farmers' problems. Among its huge product list is the Roundup Ready range, which includes GM crops.

GM crops are close to being grown commercially in the UK. There are plans to grow GM oilseed rape, sugar beet, fodder beet and maize, with GM potatoes and wheat to follow. It is possible that some of them will be grown by the end of 2001. Trials were certainly taking place in England and Wales during Spring 2001. **Figure 2** shows the location of the trials.

Figure 2 GM seed trials in England and Wales, Spring 2001

Monsanto's GM products include:

» Bollgard® Cotton: Bollgard® cotton – GM cotton, protected against insects that ruin cotton bolls. A gene from soil bacteria has been introduced, which interferes with insects' digestive systems.

» Bollgard® with Roundup Ready® Cotton: combines the insect protection of Bollgard with the ability to tolerate Roundup herbicide.

» NewLeaf® Potato: A potato with in-plant protection against Colorado potato beetle. Other NewLeaf lines provide protection against potato virus, and leafroll virus.

Herbicides include:

» Roundup®: used to control weeds on farms, golf courses and gardens.

» Roundup® D-Pak: used to control weeds in maize, cotton, rice, grain sorghum, soybeans, sugarcane, and wheat.

» Roundup Ultra: to be used with any crop with the Roundup Ready gene. It kills any plants that do not have the Roundup Ready gene.

» Roundup® Ultra RT: Roundup Ultra RT controls weeds.

The following seeds are all GM seeds, designed to be used with Roundup herbicides. They can be treated with herbicide and be unaffected – everything else is killed off except the plant.

» Roundup Ready® Maize, Cotton, Soybeans, Sugar Beets

» YieldGard® Insect Protected Maize: YieldGard® maize offers built-in protection from the European corn borer throughout the plant.

Figure 1 The *Roundup Ready* list of products from Monsanto

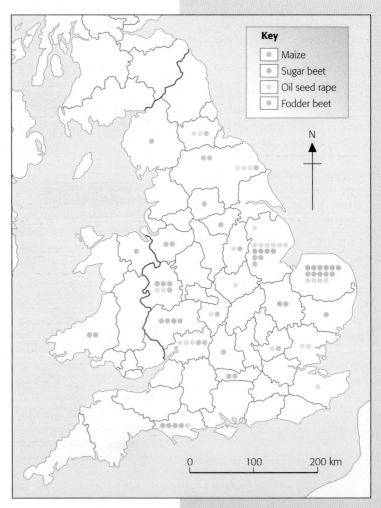

Study 2 Alternative methods of farming

GM foods are already on UK shop shelves. GM foods include tomato purée, soya and maize. Soya and maize products are used in a wide range of processed foods – 60% of processed foods contain soya products. As imports of GM soya and maize increase, more food will contain GM products.

■ What is genetic modification?

Genetic modification occurs when scientists take 'genes' from one organism and put them into another. This changes the way the organism develops, making new types of plants and animals. It allows cross-breeding between different species and organisms. For years, scientists have been able to develop plants known as **hybrids** within the same species – that is, cross breeds, in for example, cattle, flowers, grain crops. Pedigree cattle or sheep have been selectively bred for centuries by selecting particular breeding partners.

Now there are no barriers to prevent scientists taking genes from unrelated organisms (e.g. a rat and a cotton plant). Genetic modification allows genes to be crossed between organisms that could never breed naturally. A gene from a fish, for example, has been put into a tomato. Scientists cut up DNA within the genes and use the parts they want from different organisms and even make synthetic DNA.

■ What are its benefits?

Figure 3 shows the benefits of GM foods as seen by the US farming and GM industry.

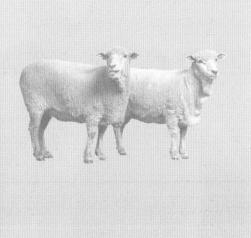

Figure 3 How they see it: the US Farming and GM industry explain the benefits

> The fruits of biotechnology will be needed to feed 2 billion more people in the next 25 years.

> Rough estimates show that 30% of cotton, 29% of soybeans, and 24% of maize acres are being planted with genetically modified seeds this spring.

> Two biotechnology companies announced that they would donate seeds of the new 'golden rice' to help prevent blindness in thousands of third-world citizens. The variety includes a Vitamin A/beta carotene component that can prevent childhood blindness found in poor, rice-diet nations. By eating this new rice, half a million people each year who would have been doomed to a life of darkness will be able to see and lead more productive lives. The new 'golden rice' is just the tip of what's coming in terms of products to improve human health, eradicate or reduce the toll of major diseases, and lengthen human life span.

Over to you

1. Using **Figure 1**, design an advertising poster to convince UK farmers that GM seeds and the Roundup range of products are worth investing in.
2. a Using **Figure 2**, describe the distribution of GM trials in the UK.
   b Why should the UK government want to keep such trials secret?
3. Draw a table with 3 columns.
   a In one column, identify all the potential benefits of GM foods from these two pages.
   b In the second, identify who stands to gain if all the benefits work out.
   c In the third, judge how good the evidence is on which such claims are made.
4. What is your view of the GM debate so far?

2.4 Genetically modified foods – the case against...

In the UK, the debate about GM foods runs deep. 77% of the UK population want them banned. In 2000, Lord Melchett, the president of Greenpeace, and over 20 others, were acquitted from a trial in which they were alleged to have damaged a crop of GM seed. The judge said that they had a right to be concerned about new seed on which there was insufficient research. In the same year, resistance in the UK and Europe forced some supermarkets not to allow GM foods on their shelves. What are the arguments against GM crops?

The current state of research

Several questions remain about GM research and crops.

- Too little is known about how genes work in most organisms. Few people know what most genes do or how they interact.
- Genetic modification is unpredictable. By inserting genes from other organisms that have never been eaten as food, new proteins are being introduced into the food chain. Will these cause allergic reaction, or affect health?
- Many GM foods contain genes which are resistant to antibiotics. Could these be passed onto bacteria in the guts of humans and animals?

Figure 1 Greenpeace leading a group of protestors at a GM trial site

Who is in control?

Genetic modification is controlled by big transnational **agro-chemical** companies, such as Monsanto in the USA. They use patent laws to 'own' every GM plant grown from their own seed. Some even force farmers to sign contracts that require the use of their own chemicals, prevent farmers from saving seed and allow the company to inspect the farm. GM engineers can even stop seed saved from one harvest from germinating the next year. This threatens the practice of saving part of a harvest to plant next year. Over 1 billion of the world's poorest people rely on saved seed. Every year, they would have to spend money on new seeds from the GM companies.

What are the environmental threats?

Herbicide Resistance. GM crops could pose threats to the environment by releasing pollen that does not react to weed-killers. GM oilseed rape, sugar beet, fodder beet, and maize are designed to be tolerant to powerful herbicides; only the crop survives the spray, and all others die. Already, many 'weeds' (i.e. wild plants) have been reduced so much by modern farming that they are now endangered, and herbicide use on GM crops could eradicate them. Wild plants in fields are important as food and habitats for insects and birds. Wipe out wild plants in the fields, and many insects and birds will be unable to feed. Many birds are already in severe decline, such as the skylark, linnet and corn bunting.

Study 2 Alternative methods of farming

Figure 2 The debate about chemical sprays – is it better to use these on ordinary seed, or use GM seeds that need no pesticide?

Figure 3 EU crop surplus store. Here crops are stored in the likelihood that they will never be used and probably be destroyed

Insect resistance. Many GM crops contain genes to make them poisonous to insects. Companies argue that these reduce the need for insecticides and benefit the environment. But not all insects are problems. Beneficial insects such as ladybirds could be affected if they eat other insects feeding on GM plants.

Genetic pollution. Many crop plants already interbreed with closely related plants. Both oilseed rape and sugar beet can cross breed with various wild plants. GM genes can be passed onto these wild plants. If GM traits such as weed-killer and insect resistance are passed to wild plants, new 'superweeds' could develop.

Crop contamination. GM and ordinary crops can cross-pollinate each other, so all farmers could find their crops contaminated. Sugar beet pollen can travel over 3 km in the wind and maize pollen several km more. Nearby organic crops could be threatened by GM pollen and cease to be organic. In the UK, test sites are supposed to have isolation zones around them to prevent this. But the zone for oilseed rape is 400 m, when its pollen can travel up to 2.5 km!

Will GM crops save the world?

Some say GM crops will help to feed growing populations by increasing yields and fighting crop diseases. However, many people in the world are suffering from **malnutrition** and hunger because they cannot afford to buy food, not because it is unavailable. There is enough food to feed everyone at the moment, yet 2 billion people are malnourished. The problem is that some parts of the world (the EU and North America) have food surpluses (**Figure 3**) and even destroy food to keep prices high, while others (many LEDCs) do not have enough.

Most GM crops now are for markets in MEDCs. Soya and maize are used mainly for animal feed and food processing in rich countries, not to feed the poor and hungry of the world. Two-thirds of GM crops at the moment are herbicide-tolerant, designed for **intensive farming** systems, with heavy use of chemicals. Many farmers in LEDCs are small, growing many different crops and they often cannot afford the chemicals needed. Another focus for research has been GM cotton and tobacco, neither of which will help to increase food supplies.

Over to you

1 a Draw a large spider diagram like the one below. On it, summarize all the arguments *against* GM foods.

b Build on the spider diagram by extending further points and issues that arise from what you have learnt.

2 Using the information on pages 210–213:
   a select what seem to you to be the strongest arguments in favour of GM foods
   b select what seem to you to be the strongest arguments against GM foods
   c which arguments seem the strongest overall? Which questions do you still feel you want answers to?

3 What is your view of the GM debate now?

FARMING

Study 3
Environmental damage and farming

Farming mismanagement can damage the environment, whether it is in MEDCs or LEDCs, and whether it is big agri-business, or individual farmers. This study focuses on the Sahel and looks at how the fragile environment is at risk, and what can be done to ensure sustainable development.

3.1 Struggling with the land – life in the Sahel

Please Father, provide the physical need of the Fulani as well: food, drinkable water, education, good health. Sometime Father, it is difficult to see evidence of your blessings among them.

This is a prayer from the Fulani people in West Africa. They are an ancient tribe of **nomadic** people whose homelands bear little resemblance to the countries of Africa. For centuries, they migrated with their cattle across areas of West Africa long before Europeans drew its current boundaries. As you read, you may find out why the prayer is important.

Figure 1 shows a photograph of a herd of traditional cattle known as zebu. Cattle herding is a traditional occupation of herders of the Fulani peoples. Migrations of cattle with their herders have occurred for centuries in an area around what is now the borders of Mali, Niger, and Burkina Faso. Until the 1970s, the Fulani rarely went south of 12 degrees latitude, because of the presence of cattle disease there. Since then, however, they have moved with their cattle to the northern areas of Nigeria, Ghana and Côte d'Ivoire. One concentration is shown in **Figure 2**, which shows a movement of Fulani herders since the 1970s from traditional grazing lands in Burkina Faso to Côte d'Ivoire. Concentrations of Fulani cattle are now highest in an area with a radius of 150 km around the city of Katiali.

The Fulani herders **migrate** seasonally with their cattle. **Figure 3** shows the climate of this part of West Africa and how this determines how they move. There are two distinct seasons, dry and rainy. During the rainy season, cattle move to new pastures. In the dry season they move closer to waterholes and river beds. 'Sleeping sickness' presents a danger to cattle and herders' cattle can die. It is carried by the tsetse fly, which lives in moist areas close to rivers and forests. It breeds during the rainy season. Its numbers rise sharply with the arrival of rain, and fall as the dry season approaches.

In recent years the number of Fulani cattle has increased sharply, at the same time as a change in climate is taking place. Vegetation and soils in semi-arid areas are vulnerable to over-grazing and over-cultivation. This is a problem that is increasingly

Figure 1 Fulani herders with their zebu cattle

Figure 2 The migration of one tribal group of Fulani people into Côte d'Ivoire from Burkina Faso

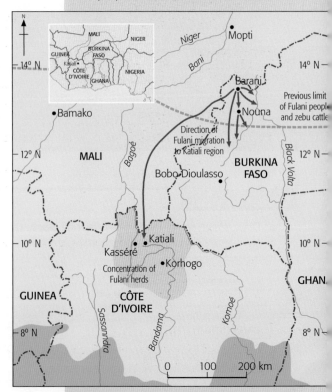

Study 3 Environmental damage and farming

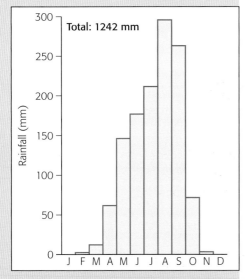

Figure 3 Rainfall in the Sahel region of West Africa

common across this part of West Africa. The region is known as the **Sahel** – it means the 'transition zone' between desert and grassland. Like others living in the dry areas of West Africa, the Fulani people have problems in finding sufficient grassland and water. You can find out more about the Sahel on pages 216–221.

Poverty in Burkina Faso

Pressures on land in Burkina Faso and declining income have prompted the Fulani and other nomadic herdspeople to settle on the land and to farm. Their sales of beef at markets have faced stiff competition from Latin America and from beef imported from subsidized European Union producers.

The rural regions of Burkina Faso are some of the poorest in Africa. Population pressures have damaged the landscape and made the struggle to eke out a meagre existence worse. Many people from rural regions migrate, either to towns and cities within Burkina Faso or to neighbouring countries, particularly Côte d'Ivoire. Wages that migrants send home are an essential part of many household incomes. Although 90% of the Burkina Faso's population earn a living of sorts from farming, the majority farm only at the **subsistence** level – that is, they make just enough for themselves and their families.

Some reasons for low crop yields are:
- most farming is on small plots which produce just enough food for families
- nutrition levels are low, so the workers are not strong
- most agricultural labour is done by women
- as the land loses its fertility through intensive use, so food yields fall
- most farmers use traditional tools and methods that limit crop yields
- use of fertilizer is minimal, keeping crop yields low.

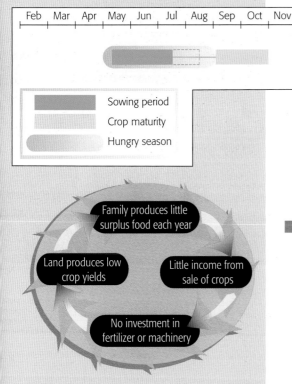

Figure 4 The rhythm of the year and the hungry season in Burkina Faso

Figure 5 The vicious circle of poverty in the Sahel

The hungry season in the Sahel

The hungry season occurs between the beginning of field preparation and the harvest (**Figure 4**), when food supplies from the previous harvest are at their lowest and energy demands are highest. Its length depends on the quality of the current and previous harvests, but it typically ends shortly before the cereal harvest, as cereal prices fall and early crops and gathered foods become available. In Burkina Faso, grassland and water are becoming very scarce. Most herders are concerned about whether they and their animals will be able to survive under such conditions before each rainy season begins in April.

Over to you

1 Make a sketch copy of the graph in **Figure 3**. Label it to show:
   a when grass growth is likely to be rapid
   b when newest and best-tasting grasses are available for cattle to feed
   c periods of drought
   d when grasses are likely to be tough and woody and difficult for cattle to eat.

2 Compare your labels with **Figure 4**. How can the hungriest time be the wettest?

3 Make a large copy of **Figure 5**. On it label features of:
   a the traditional Fulani way of life
   b their lifestyles that explain why they are trapped in poverty.

4 Is damage to the environment in this area caused by people or by natural processes? Explain.

FARMING

3.2 Farming and drought in the Sahel

Every so often, a drought takes place which shocks the world. Comic Relief in 2001 focused upon Burkina Faso, a country which has had several droughts in the past 30 years. Further east, during 2000, droughts in Eritrea led to millions of dollars being given for aid programmes. These were not isolated events, but are part of a pattern of climate change since the 1960s, which has affected a broad zone known as the Sahel, shown in **Figure 1**.

The Sahel lies on the margins of the Sahara Desert, and is a transition between tropical hot desert to the north and grasslands to the south. It measures about 3000 km long and 700 km wide. It depends on the monsoon for its rainfall, but rainfall is variable. **Figure 3** shows some aspects of this problem. This affects the lives of the Fulani and others.

- Trees die, made worse by overuse of fuelwood.
- Grazing land is under pressure from so many cattle, and when it is not regenerated by summer rain.
- As populations grow here, land use becomes more intensive.

For the Fulani, lower rainfall puts pressure on a traditional way of life. Their lifestyles become focused upon finding water, usually around drinking wells (**Figure 2**). These are based around **boreholes** that pump water from below ground, and which provide stopping points for cattle herdspeople. However, they cause a further problem as they tend to attract large numbers of people and their cattle, placing huge pressure on the grassland and semi-desert environment.

What caused the problems?

The last 30 years of drought and famine in the Sahel have happened due to the longest and most dramatic change in rainfall in any one region on the Earth since reliable weather readings began. Three possible causes may be responsible for this drought:

Land degradation within Africa (**Figure 4**). Some believe that degradation in the Sahel is not the result of climate change, but of pressures brought about by poverty, as people search for food and fuelwood. Increasing pressure brought

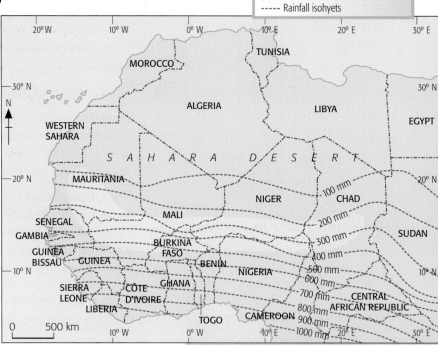

Figure 1 The Sahel Region

Figure 2 A well in the Sahel

Figure 3 Rainfall in the Sahel. The graph takes the 1961–90 period and averages it. The 0 line is average. Anything above the line shows wetter years, anything below shows drier periods

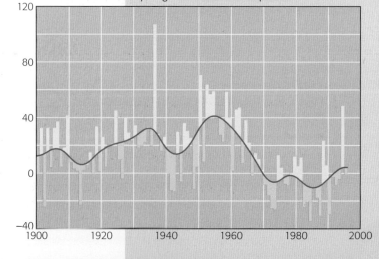

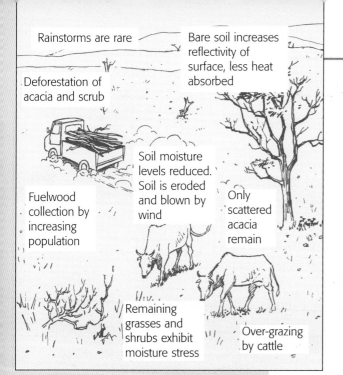

Figure 4 Land degradation

Figure 5 The effects of drought in three countries in the Sahel

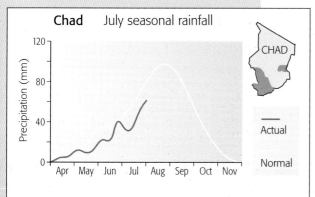

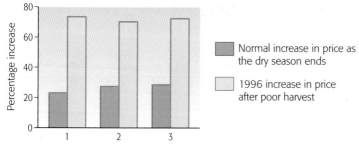

Study 3 Environmental damage and farming

by cattle and farmers would also put pressure on plants which, faced with drought, would gradually die off, exposing soil beneath to removal by wind. With fewer trees or plants, less moisture would be retained in the soil, fewer trees and plants would retain moisture, and therefore there would be less moisture to form rain in the atmosphere.

Natural climatic variability. No two years are alike anywhere, and it is natural that rainfall should vary between one year and another. This, however, would not explain longer-term trends in the Sahel towards lower rainfall.

Climatic change linked to global warming. This is theoretical, but it is possible that warming seas and atmosphere alter the flow of winds that bring rain during the wet season. Such disturbance might prevent rains getting very far inland towards the Sahel in one year, while encouraging them the following year. Increasing global warming might also explain why the trend of rainfall is downwards.

How are people affected by drought?

Figure 5 shows three effects of drought based on what happened in 1996. **Figure 3** shows how 1995 had been an exceptionally good year for rain, and in this year crops and grass grew well. **Figure 5** shows how, as rainfall fell below average in 1996, three countries – Chad, Niger and Mauritania – were affected in different ways.

- In Chad, rainfall fell only in the red area shaded. The graphs show that it was below the normal (or 'average').
- In Niger, millet (the staple food crop) prices rose sharply.
- In Mauritania, sorghum (the staple food there) rose in price sharply at five different market towns.

Over to you

1 Draw a sketch map to show the Sahel in West Africa. Shade in the Sahel, add a key to your map, and label the following:
  a which countries are included in the Sahel
  b effects on different countries and people as shown in this study.
2 Explain why wells (**Figure 2**) can cause problems as well as solving them.
3 Describe the pattern of rainfall in the Sahel using **Figure 3**. Use the data in the graph to assess whether you think the climate really is changing in the long-term.
4 Draw a diagram to show the knock-on effects of how land degradation can lead to lower rainfall.
5 Would you say the problems faced by the Sahel are caused by human activity or by natural causes? Explain.

FARMING

3.3 Giving the Sahel a future

One of the issues in improving life for people in the Sahel is that different people have different views about how it can be done. Here, three projects are described – you decide which you think is best. Some are in Mali (**Figure 1**) while others are elsewhere in the Sahel.

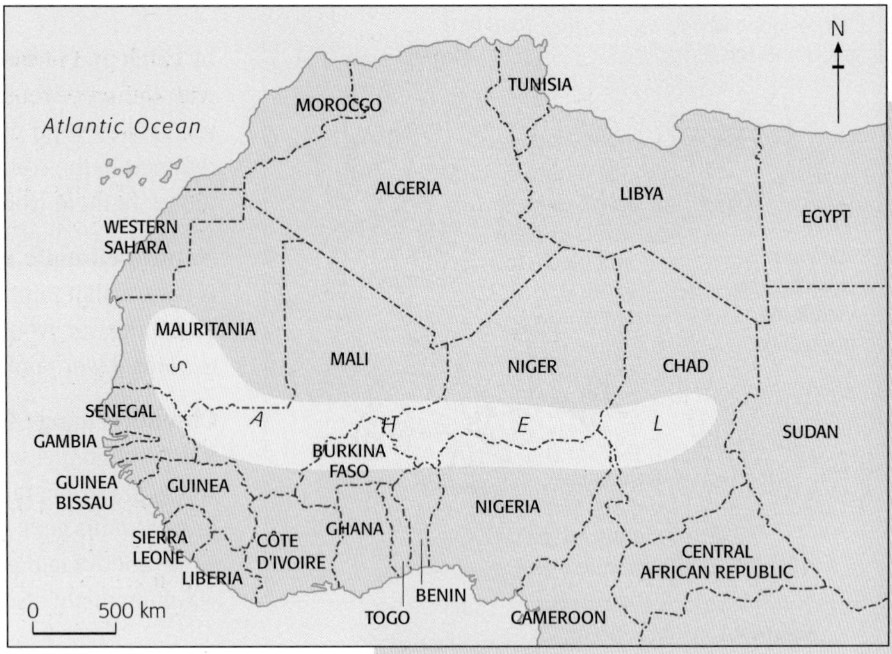

Figure 1 Mali and other Sahel countries

Using irrigation to increase crop yields

With so many problems to face, it has been difficult to encourage Sahelian farmers to raise crop yields. In the south, where rainfall is greater, farmers can use **irrigation**, using water stored from the rainy season. Four ways have been used to increase crop yields.

- Adopting new seed varieties, fertilizers, animal traction, and small-scale irrigation technologies. Farm research has contributed to increases in maize yields in parts of Mali, Senegal, and Burkina Faso by breeding varieties that are resistant to drought.

- When Malian farmers applied fertilizer to the new varieties, they tripled maize yields, and in turn planted 10 times more land with maize.

- Improved millet farming and cotton cultivation is being adopted in Senegal, used in conjunction with chemical fertilizers and pesticides. These are increasing crop yields in Burkina Faso and Mali. In Mali, the use of animals to cultivate heavier, richer soils has helped farmers to increase yields further. Small ditches constructed in the depressions between dunes alongside Lake Chad and in seasonal riverbeds have provided enough moisture to double wheat yields.

- An irrigation scheme in Niger in the 1990s resulted in a tripling of rice yields. This was due to technical factors such as better water management, training, a better water supply system, and increased farm mechanization. Flooding in 1994 caused a temporary setback, but rice farmers continued to improve yields.

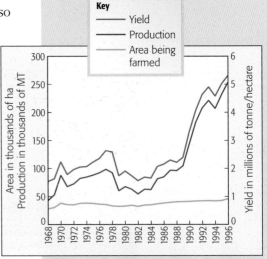

Figure 2 Rice yields in Mali increased following the use of irrigation techniques

Seed cross-breeding

Another solution comes from Mali, where research has found the best strains of millet for drought conditions. Researchers selected the most productive varieties among the traditional species and set about trying to cross-breed **hybrids**. This is not an example of GM

Study 3 Environmental damage and farming

technology, but simply using existing strains to develop a hybrid that can withstand drought. A seed bank was set up with all local and some wild millet varieties, about 1200 in all.

Early-maturing varieties from southern Mali (where the growing period is short) were selected and planted in the north of the country. Even here, where rainfall is low, good harvests were obtained. In the southern part of the millet-growing region, small farmers plant parts of their fields with the early-maturing variety. By harvesting their own early millet and feeding their families, they are harvesting at a time when they would normally have to pay high prices for millet from dealers.

One new hybrid increased yields by up to 13% or 60 kg per hectare to a total of 514 kg per hectare. Its growing period is one week shorter and it is more drought and pest-resistant. 60 kg of millet per hectare may seem a small increase, but it is enough to feed one child for one year.

■ The Green Cross programme in Burkina Faso

Figure 3 Planting seedlings in the Sahel. The raised earth traps water

Green Cross is an international environmental organization, whose president is the former USSR president, Mikhail Gorbachev. A tree-planting project aims at stopping the progress of the desert in several provinces of Burkina Faso, by planting over 370 000 trees adapted to the different parts of the country. Green Cross Burkina Faso has launched a programme for fighting desertification that will last 5 years and includes:

- the creation of nurseries of young plants
- tree planting by local people
- teaching people about plant growth
- developing solutions for better use of fuelwood on fuel-saving stoves.

Experience acquired during the project will be used to expand it to other regions of Burkina Faso as well as other parts of the Sahel.

Other similar projects exist, such as Tree Aid, a British charity formed through the Royal Forestry Society, which has produced 2 704 935 seedlings to plant 96 km of shelterbelts, orchards, woodlots, trees around fields and homesteads, and has supported 53 projects in 14 countries in Africa.

Over to you

1. On a sketch map of the Sahel (drawn from **Figure 1**) identify and label each of the projects mentioned on these two pages.
2. Take each of the three projects in turn. For each one:
   a. describe its aims, and its methods
   b. describe its advantages and disadvantages
   c. say how successful you think it has been.
3. Which project would you say has greatest economic potential? Why?

FARMING

3.4 Improving life for the Sahel's poorest

Sahelian countries are desperately poor. **Figure 1** shows the extent of poverty in Burkina Faso, the country in focus on these two pages. How can the country be helped?

The projects mentioned on the previous two pages are based upon financial aid – perhaps from Europe or the USA. However successful these are, projects that start with local people sometimes have even greater success. On these two pages, projects are described which attempt to improve lives for people living in the Sahel. Although they involve people from overseas, the solutions are less costly and use local resources – including people!

Siguin Voussé is a village of about 50 scattered farmsteads in eastern Burkina Faso, West Africa. Rains have become poorer in recent years. As a result, millet and sorghum harvests have decreased, leaving families with no surplus food to sell, and no cash to pay for clothes, medicines, and schooling. In 1995, community leaders in this and other local villages formed a group called Dakupa, and things began to change. Dakupa is involved in the following three projects.

■ Providing loans for women

Maimunata Nombre, a Dakupa volunteer, supports women's groups (**Figure 2**) in several villages, and many women now take out an annual loan and use it for small-scale business. Awa Bundani explains how she used her loan: 'I bought two sacks of peanuts for £20, and should be able to sell them for at least £25 later in the year. I will also use some for making croua croua [peanut rings] to sell.'

■ Diguettes and water-harvesting

The dry northern region of Burkina Faso has been affected badly by drought. Deforestation and over-use of land left much of the province unable to support subsistence farming.

In 1979 Oxfam began to support a project called Projet Agro-Forestier. Over the years, most of the trees and grasses in the area had been cleared for farming, and **topsoil** was washed away when it rained. The project aimed to prevent further soil erosion, and to preserve as much of the rainfall as possible. It encouraged local farmers to build **diguettes**, small barriers of rocks along the contours of the ground. The diguettes, which follow the contours of the gently sloping fields, hold back rain-water so that it soaks into the ground, and any soil particles are trapped. It has been a great success. Crop yields have increased and the farmers are more self-sufficient. Over 400

Burkina Faso	
Life expectancy at birth	46
Population (millions)	10.9
Population growth rate %	2.8
Per capita GDP US$	230
Education	
Primary enrolment – girls (%)	24
Primary enrolment – boys (%)	37
Pupil–teacher ratio in primary school	54
Secondary school enrolment (%)	7
Adult literacy – women (%) (1995)	9
Adult literacy – men (%) (1995)	30
Health	
Health expenditure per capita (1995 PPP $)	43
Maternal mortality (per 100 000 live births)	930
Infant mortality (per 1000 live births)	110
Under-5 mortality (per 1000 live births)	169
Number of doctors per 100 000 inhabitants	3
% of malnourished children under 5	30

Figure 1 Socio-economic data for Burkina Faso

Figure 2 Future businesswomen: a women's group in Burkina Faso

Study 3 Environmental damage and farming

villages have built diguettes. In one test done by farmers, soil depth in an unprotected area decreased by 15 cm while in a diguette area soil depth rose by 18 cm.

Three elderly women explain how their land has been damaged:

'We have cut down all our trees and as a result the wind makes us suffer because it erodes the soil. We have been educated about this problem and now we build diguettes to try to reduce the damaging effects of erosion. Ravines can also damage our fields and to stop these enlarging we put stones in the path where the water flows. We remember when, to find firewood, we only had to look behind our houses; today we have to walk up to 8 or 9 kilometres.'

Figure 3 Making traditional baskets, a local craft in Burkina Faso

■ Education

Awa Bundani has a daughter – Mariam. School fees and the cost of Mariam's books amount to only a few pounds a year. But, in a country where only two out of five children attend primary school, the family's improved harvests and Awa's new income have helped to guarantee Mariam an education. This is important in distant rural communities where education may not be valued.

A nomad-parent explains: *'When I send my child away for six months to learn how to tend the cows, he will return with useful knowledge that can help our family survive, and a little cow as his wage for the assistance he has given. What will my child return with if I send him away to school for six years?'*

Mariam means to make the most of her opportunity: *'I'd like to become a nurse, because there isn't one in Siguin Voussé.'*

Figure 4 Building a diguette, or stone line, during the dry season

Sustainable development
◆ is long lasting
◆ helps rather than damages the environment
◆ makes use of local resources and people's skills
◆ may actually improve the environment

Over to you

1. Select five sets of data in **Figure 1** that you think show Burkina Faso's poverty most clearly. Justify your selection.
2. Consider each of the three projects set up by Dakupa with Oxfam.
   a. Why should loans for women (rather than men) be a priority in Siguin Voussé?
   b. Show how diguettes work using labelled diagrams.
   c. Explain to the nomad-parent (in the section on Education) whether you feel that six years of schooling is worthwhile, and why.
3. Compare the three projects on these pages with those on the previous two pages. Select two that you feel are most worthwhile. Justify your choice.
4. Study the definition of sustainable development in the box above. Do you think Oxfam's work with Dakupa is sustainable? Give your reasons.

RECREATION AND TOURISM

Study 1
Using the countryside for recreation

More and more people are visiting the countryside. This study looks at the reasons for this, and how different landscapes can provide different opportunities for recreational tourism. We look at how we can protect the countryside, and at one National Park in particular, the Lake District.

1.1 Changing holidays

It may not always feel like it, but people in Britain now get more leisure time than any previous generations. Of course, this varies from one person to another. People who are self-employed often have to work long hours to keep their business going, and holidays for them mean times when they cannot earn money. But for many people, leisure time, and how they spend it, drives one of the biggest industries in the world – **tourism**.

Consider **Figure 1**, which tells the story of someone of your age in the 1950s – now your grandparent's generation. The questions they were asked were designed to find out what leisure time people had then, and how they spent it. Leisure time spent away from work is usually known as recreation. Travel away from home – for a day, weekend or longer – is known as tourism. This section is all about how we spend our time and where, and the effects it has on other people.

Q1	When you were growing up in Essex in the 1950s, how much holiday did you have with your family?
A	My father got 2 weeks' holiday a year.
Q2	Where did you go on holiday and how?
A	For the first few years of my life, we went to Clacton in Essex, about 20 miles away. We used to cycle or get the bus. As I got older, we used to go to Cornwall by train.

Q3	Did you get any other holidays in the year? Or any days' out?
A	No. My father only had one day a week off. We had no car as my parents couldn't afford it. We used to go out for a few odd days locally, and I used to cycle to see my grandparents and aunt and uncle. Once or twice a year we would go to London for the day – it was only 40 minutes on the train.

Figure 1 Leisure in the 1950s: how is it different for you?

Study 1 Using the countryside for recreation

Q1	How much holiday do you have with your family each year?
A	We don't have one each year. In my family, we've never done what many other British people do and have a yearly holiday.
Q2	What do you do for holidays then?
A	My family is originally from India, in the Punjab region. So we save up for 3–4 years and have a long holiday there to see our family. We usually go for 6–8 weeks at a time to make the most of it.

Q3	Don't you get any time away apart from that?
A	We spend a lot of time as a family. My relatives live close by and we spend time with them. We did go to Cuba two years ago.

Figure 2 Our kind of holiday

Year	Weekly average household income
1970	£252
1980	£281
1990	£340
1995–6	£340
1999–2000	£391

Note – these figures are based on current prices and remove the effects of inflation, so that data can be directly compared between one year and another

Figure 3 Average income per week for households in the UK

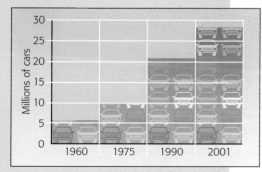

Figure 4 Car ownership in the UK

■ Why is there more leisure time now?

How much leisure time you get, and when you take it, depends on who you are.

a) Families with children of school age often try to take time off together at holidays or half-terms.

b) Many people in their 50s and 60s have taken early retirement on a pension. They have access to recreation and can travel away at any time. Their money and time is sought after by many tourist companies chasing the 'grey pound'.

Many full-time working people are now entitled to four weeks' paid holiday a year. Some have far more than this. Working hours, however long they may seem, have got shorter since your grandparent's generation, when six-day working weeks were common.

Of course, leisure time varies, as does the way we spend it – as shown in **Figure 2**. Some families are fortunate in earning high salaries that enable them to buy second homes or travel on long-haul holidays. Others are forced by low wages and difficult working hours into lives with little leisure time. But generally, British people have more time, with more money to spend than before.

Over to you

1 Using questions like those in **Figure 1**, ask three different age-groups how they spent their leisure time when they were growing up. Talk to:
  a your grandparents' generation   b your parents' generation
  c your own generation.

2 What differences do you notice? Write up your results on a large copy of the following table.

	Grandparents	Parents	Your generation
Q1			
Q2			
Q3			

3 Explain the differences you notice.

4 How might the ways people spend their holidays and leisure time vary with:
  a household income?
  b social factors such as family background?
  c increasing car ownership?

RECREATION AND TOURISM

1.2 To the hills!

How do you view the countryside? Perhaps you live in a rural area, and experience long periods of quiet in winter, only to be invaded by tourists in summer. For years, people in Cornwall have referred to tourists as 'emmetts', as though they are a different race. Perhaps you live in an urban area and are one of the 'emmetts' who head for the countryside for days out, weekends or holidays. Perhaps you go for the 'Sunday drive' with family or friends, or on an occasional geography field trip. Whichever way, the number of visitors to rural areas is higher now than ever before (**Figure 1**). It is difficult to know exactly how many people visit the countryside, but one survey in the Lake District gives us a clue.

Figure 1 The Lake District: tourists flock ...

- 3941 people were asked about their most recent visit to the Lake District at roadside surveys and at sites such as car parks and visitor attractions.
- The results were collected and compared with other similar surveys in other national parks in the UK.
- As a result, estimates were made of the numbers of visits made each year to the Lake District and other national parks. In the mid-1990's, about 17 million visits were made.
- Some visitors in the survey were day trippers, while others were staying for a longer period. An estimate was made that 22 million visitor days were spent in the Lake District. In this case, one visitor staying for one day equals one visitor day.

Figure 2 ... to see scenery such as this

Study 1 Using the countryside for recreation

■ So what's the attraction?

The Lake District has been enjoyed by tourists since Wordsworth made it famous with his poetry in the 19th century. Railways took people to other rural areas too – from London to Surrey or Sussex, from Leeds into the Pennines, from London and Birmingham to Cornwall and Devon (**Figure 3**).

Now, different rural areas have a range of attractions. For example, the Camel cycle trail in north Cornwall runs along an old railway line.

National Trust membership has risen from 2 million in 1990 to 2.7 million in 2001. The Trust uses membership fees, grants and bequests to conserve and protect buildings and land from further development.

There are all sorts of attractions in rural areas. There are theme parks, like Alton Towers in Staffordshire (**Figure 4**). There are working farms across the UK. There are even vineyards, but mainly in southern Britain! Some attractions are targeted at families, and others at older age groups.

Most visitors to these attractions are day-trippers. The number of visitors from long distances away is usually less than the number from close by. This is known in geography as '**distance decay**'.

Figure 3 Tourists first saw Britain by train

Figure 4 A grand day out: at Alton Towers...

Figure 5 ...or at a historic house such as Lanhydrock House, Cornwall

Over to you

1 Why should the number of visits made by British people to rural areas have increased?

2 **a** In your class, which rural areas have been visited most recently:
   i) for day trips   ii) for longer breaks?

   b Plot the location of your school on a map of the UK, and then plot the places visited on the map.

   c Draw circles centred on your school of radius 100 km, 200 km, 300 km and 400 km on your map.

   d Count the visits inside each circle, and draw a graph to illustrate the result.

   e Is there a 'distance decay' – i.e. the further away, the smaller the number of visits?

3 In 300 words, summarize the variety and location of rural landscapes and attractions in Britain.

RECREATION AND TOURISM

1.3 Protecting the countryside

In 1997 Tony Blair's new Labour government set out to tackle the issue of the 'right to roam'. There was to be a series of laws designed to open up the paths and tracks of the countryside to visitors, on which people had been prevented from walking by landowners and companies. In fact, this was part of a much longer-term change, which began in the 1930s. Then, ramblers and visitors from towns and cities all over Britain took part in a protest that led to measures to open up the countryside, and also to protect it for future use. Some of the areas protected as a result were known as **National Parks** (**Figures 1–4**).

■ What are National Parks?

There are now eleven national parks in England and Wales (**Figure 4**) which are protected by law from certain kinds of development. Each has its own authority which can make recommendations about future development such as building. But the authorities do not own the land, unlike most national parks in the USA, Canada or Australia. Private landowners own almost all the land, and they do not always agree with what the authorities say. This leads to **conflict.**

Figure 1 What features do the Yorkshire Dales...

■ What's special about the National Parks?

Figures 1–3 show the different kinds of landscapes in the national parks. Study the location of these, using **Figure 4**. They have been made national parks because of their:

- landscape quality
- height (or relief)
- landforms, such as the valleys of the Yorkshire Dales (**Figure 1**)
- ecology (or plant and wildlife), as in the New Forest (**Figure 2**)
- geology (or rock type), such as the 'tors', or exposed rocks, of Dartmoor (**Figure 3**).

Figure 2 ...have in common with the New Forest...

Figure 3 ...and Dartmoor?

Study 1 Using the countryside for recreation

These areas face different pressures because different people want to use them in different ways. Tourists are attracted to all of them, but for different reasons. Consider who might want to visit these parks and why. Try also to imagine where these tourists would be likely to travel from (look at **Figure 4**).

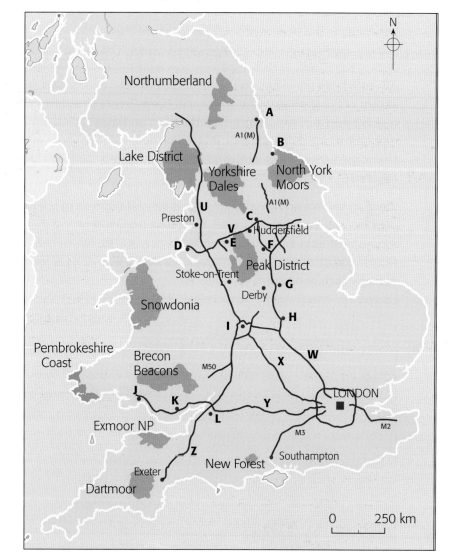

Figure 4 The National Parks of England and Wales. How do you think the cities and motorways shown affect visitor numbers?

Over to you

1 What do you think is special about landscapes like those shown in **Figures 1**, **2** and **3**?

2 Study **Figures 1–3**. Make a copy of the diagram below and add as many features as you can that make the National Park landscapes unique. Use examples from the photos, or from National Parks you know.

3 a Study **Figure 4**. Draw your own map of the National Parks of England and Wales.

b Using an atlas, identify the cities A–L and motorways U–Z shown. Label these on your map.

4 Which National Parks have cities closest, and would therefore be under most pressure from visitors?

5 In 200 words, say

a why some people might see certain landscapes as needing protection from development

b whether you think they need protection and why.

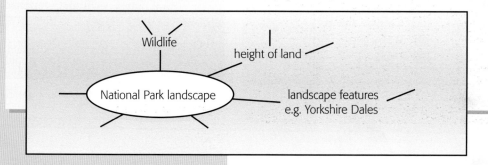

RECREATION AND TOURISM

1.4 The Lake District landscape

The Lake District is one of Britain's most popular national parks, and attracts millions of visitors throughout the year. It is one of England's most dramatic landscapes with some of its highest peaks, such as Helvellyn (**Figure 1**) at 950 metres above sea level. How has this landscape emerged, and what pressures is it facing?

■ How does geology affect the landscape?

All landscapes are influenced by the rocks that have formed them – known as the **geology** of the landscape. **Figure 2** shows the different rock types in the Lake District. Find Helvellyn on **Figure 5**, then the rock type of which it is formed. Then compare it with the photo of the landscape between Kendal and Windermere (**Figure 3**) and Skiddaw (**Figure 4**).

These varied landscapes offer the opportunity for all kinds of outdoor activities and for study. Geologists, naturalists, and those with a serious interest in the landscape and its history will be drawn here. But most of all, they offer opportunities for sightseeing or more active pursuits, such as hiking or climbing.

Figure 1 Different rocks, different views: from the high peak of Helvellyn...

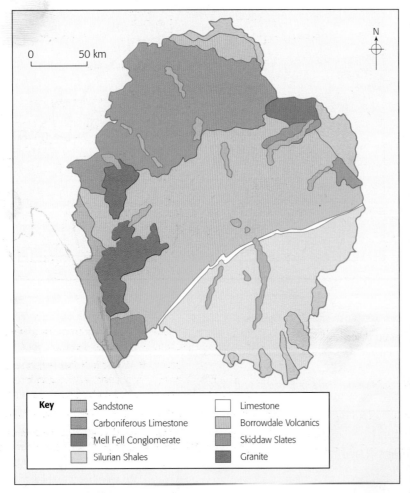

Figure 2 The geology of the Lake District

Key:
- Sandstone
- Carboniferous Limestone
- Mell Fell Conglomerate
- Silurian Shales
- Limestone
- Borrowdale Volcanics
- Skiddaw Slates
- Granite

Figure 3 ...to the rolling scenery around Kendal...

Study 1 Using the countryside for recreation

How accessible is the Lake District?

Figure 5 shows the key features of the Lake District, and how it is so **accessible** for people from all parts of Britain. Compare this map with an atlas map of the UK, and identify which cities lie within easy reach of the Lake District. As one of the more accessible parks it is also one of the busiest, and this puts pressure on the landscape.

Figure 4 ... to Skiddaw in the northern Lake District

Figure 5 The Lake District and how people get there

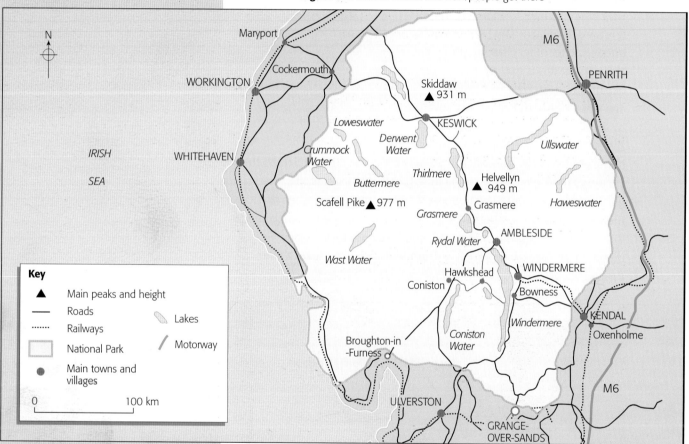

Over to you

1 a Draw a large sketch copy of **Figure 2**, showing the geology of the Lake District.

 b On it, add labels to show:
   i) the location of Helvellyn, Skiddaw and Scafell (use **Figure 5** to help you)
   ii) a description of the landscape for the two rock types for Helvellyn and Skiddaw.

2 How is the appearance of the landscape and its height linked to the rock types? Give examples.

3 Consider the following outdoor activities:

fell walking, rock climbing, camping.

For each one, add a label to your map from **1a** to show where it is most likely to be found and why.

4 The Lake District National Park Authority has asked you to place adverts in evening newspapers in all the cities within 2 hours' drive of the Lake District to advertise its attractions.

 a Use an atlas to identify in which cities you would advertise.

 b Design an eye-catching advert targeted at people of your age who might never have been to the Lake District before.

RECREATION AND TOURISM

1.5 A place for everyone?

As well as geology, several other processes have made the Lake District landscape what it is today. One of the problems facing planners there is that the landscape has evolved over thousands of years, yet could be ruined quickly if development is uncontrolled.

■ The Lakes - a relic of a past climate

It is hard to imagine the Lake District under a sheet of ice, like the photo in **Figure 1**. Yet this is exactly what it was like between 20 000 and 10 000 years ago when northern Britain was covered by ice sheets and highland **glaciers**. Just as the mountains of Greenland are being eroded today beneath the glaciers shown in **Figure 1**, so too the valleys of the Lake District were eroded. In the valley bottoms, the glaciers 'scooped' out huge hollows, which formed lakes when the ice melted. The débris left behind when the glaciers melted often formed dams of boulders, sand, and gravel, know as 'moraines', which dammed the lakes. Lakes such as Thirlmere shown in **Figure 2** were formed like this.

Figure 1 20 000 to 10 000 years ago the Lake District landscape was like parts of Greenland today: covered with ice!

Shape of valley before glaciation

Figure 2 Thirlmere. The dotted line shows the shape of the valley before the Ice Age

Study 1 Using the countryside for recreation

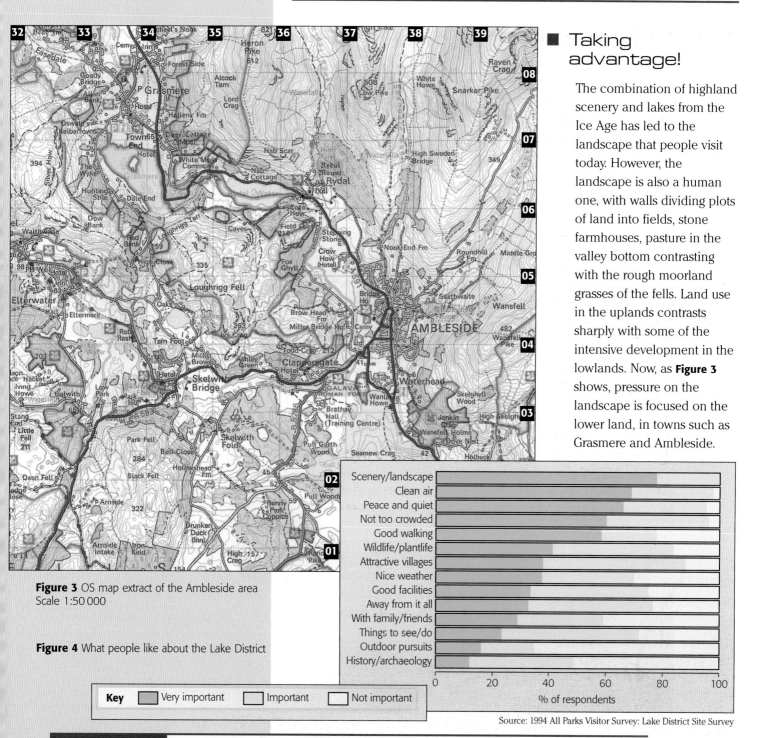

Figure 3 OS map extract of the Ambleside area
Scale 1:50 000

Figure 4 What people like about the Lake District

Key: Very important | Important | Not important

Source: 1994 All Parks Visitor Survey: Lake District Site Survey

Taking advantage!

The combination of highland scenery and lakes from the Ice Age has led to the landscape that people visit today. However, the landscape is also a human one, with walls dividing plots of land into fields, stone farmhouses, pasture in the valley bottom contrasting with the rough moorland grasses of the fells. Land use in the uplands contrasts sharply with some of the intensive development in the lowlands. Now, as **Figure 3** shows, pressure on the landscape is focused on the lower land, in towns such as Grasmere and Ambleside.

Over to you

1. How can scientists be fairly sure that the landscape of the Lake District was formed during the Ice Age, and not by other processes?

2. On the Ordnance Survey map in **Figure 3**, identify and give 6-figure grid references for:
   a. lakes formed by glaciers
   b. a dam or 'moraine' at the end of a lake.

3. Give names, 4-figure references, and heights above sea level for:
   a. four villages
   b. three farms
   c. one youth hostel
   d. four campsites.
   Why should all of these be found on lower ground?

4. Ambleside is the largest settlement shown on the map. Suggest:
   a. what shops you would expect to find in Elterwater and Grasmere, and why
   b. what *additional* shops you would expect to find in Ambleside.

5. Study **Figure 4**. Select any 5 reasons why people like the Lake District, and produce a 2-sided visitor pamphlet to show where the things they like can be found in the Ambleside area of **Figure 3**.

RECREATION AND TOURISM

Study 2
The impact of visitors on the countryside

Visitors to the countryside bring benefits, but can also cause problems. This study looks at some of these issues, and also considers how tourism could develop in the future.

2.1 What are the benefits of tourism?

Tourism undoubtedly brings benefits to the Lake District. This was never more clearly shown than in the first half of 2001, when foot-and-mouth disease occurred in Cumbria, the county in which the Lake District National Park is located. As a result of the disease, footpaths were closed, and government ministers tried to persuade the public not to go to the countryside. For the Lake District, this was a (fortunately temporary) problem for the tourist industry, which suffered badly.

The Lake District National Park local plan has found that tourism is vital to the region, as shown in **Figure 1**. It helps farmers who, even before the foot-and-mouth outbreak, were doing badly.

- 31 000 jobs in Cumbria as a whole arise directly from tourism
- This accounts for 15% of Cumbria's workforce
- Most businesses are small-scale, and in many cases family-run
- Within the Lake District National Park, 33% of the workforce is employed in hotels, catering and distribution compared with 6% in the UK as a whole
- In some parts of the Lake District tourism is even more important, with 55% of jobs in Windermere and Keswick arising from it
- Tourism supports many other services such as bus and rail networks, village shops, pubs, and farms

Figure 1 Why tourism matters... the facts

Figure 2 Why are these caravans in the Lake District so important to the area?

Study 2 The impact of visitors on the countryside

Type of accommodation in the Park	Holidaymakers staying in the Park % of respondents
SERVICED	
Hotel/Motel	17
Guest House	5
Pub/Inn	1
Bed & Breakfast	13
Farmhouse (B&B)	2
Outdoor activity/Training centre	1
Total	39
SELF-CATERING	
Caravan (static)	8
Caravan (touring)	10
Caravanette/Camper-van	2
Camping (camp site)	10
Camping (open country)	2
Self-catering accommodation	19
Timeshare	1
Youth hostel	2
Bunkhouse barn	*
Climbing hut	*
Second home	2
Boat (own)	*
Wooden chalet/Log cabin	1
Total	57
Homes of friends/relatives	4
Other	*
Base Number	(1420)

Figure 3 Where people stay in the Lake District
Source: 1994 All Parks Visitor Survey: Lake District Site Survey

Figure 3 shows the variety of accommodation used by visitors to the Lake District. Wherever they stay though, tourists bring other benefits.

- Local tourist amenities such as the Brockhole Visitor Centre (**Figure 4**), run by the Lake District National Park, has over 200 000 visitors each year. The centre not only promotes the Lake District but has an educational purpose, and accepts school groups.

- Local services, such as car parks run by the National Park Authority and by local councils, raise significant amounts of revenue.

- Most visitors – over 80% – are repeat visitors who return to the Lake District most years. However, such visitors need variety as well as being able to see the area they enjoy so much. Much is being done now to extend the range of activities and amenities to involve young people.

Figure 4 Visitor Centres like this one at Brockhole help people learn about and appreciate the Lake District.

Over to you

1. Read **Figure 1**. Use the heading 'The importance of tourism in the Lake District' and:
   a. select any three examples of data to show how tourism is important, and draw graphs to illustrate it
   b. describe how bus and rail networks benefit from tourism
   c. describe what might happen to village shops and pubs without tourism.
2. Suggest several ways in which farms can benefit from tourism.
3. Draw a pie-chart to show the proportion of people staying in the following: 'Serviced', 'Self-Catering' accommodation, and 'Homes of friends and relatives'.
4. a. Outline the economic advantages for the Lake District of:
   i) serviced accommodation
   ii) self-catering accommodation.
   b. Are there any disadvantages?
5. How could more variety be introduced into the Lake District, without spoiling what visitors come to see? Consider:
   a. how teenagers might be encouraged and the amenities they would want
   b. how variety could be provided for active, older people
   c. how large companies or corporations might be attracted to use the Lake District for conferences.

RECREATION AND TOURISM

2.2 What problems does tourism cause?

Not all the effects of tourism are beneficial. So many people now visit the Lake District National Park that some negative impacts are bound to be felt, especially in the summer months. These are most felt in tourist '**honey-pots**' – places that attract most people. Some of these problems are economic, others environmental.

■ Traffic congestion

Roads in the Lake District National Park are mainly narrow, bordered by stone walls. In winter, traffic levels are low, and roads fairly empty. But in summer, the situation changes. Congestion occurs:

- at tourist 'honey-pots', such as Windermere, especially at the entrances to car parks
- at bottlenecks, such as Ambleside, where roads converge
- on roads where coaches have to negotiate narrow and sharp bends
- on public holidays, where day trippers combine with long-stay tourists.

As a result of the traffic congestion, local people are unable to do their shopping or get to or from work easily. Events such as road traffic accidents or medical emergencies become crises when services are unable to get through the traffic.

But some tourists are less problematic than others. Visitors staying for some time spend more than those on day visits. They shop more and spend more each time they shop. They travel less each day, and take short rather than long journeys.

■ High house prices

The large number of tourists has led both to the growth in housing used for summer letting, and in holiday homes bought by wealthier people in cities such as Leeds or Manchester. For those who can afford to buy, high rents in summer months make old cottages and barn conversions attractive for investors. Because of the demand for property, many local people cannot afford to buy houses.

Those priced out of the market tend to be:

- workers in the tourist industry, shop workers, and farm labourers, who are among the lowest-paid wage earners in the UK
- young people, attempting to rent or buy property when they first begin work, and especially young families
- the elderly, or others on restricted or low incomes.

Figure 1 Footpaths as well as roads in the Lake District get congested!

Figure 2 How people travel to the Lake District National Park

MODE OF TRANSPORT	All Visitors: % of people (1) from sample of 1508
PRIVATE	
Car/Van	85
Landrover/4 wheel drive	<1
Camper van	2
Motorcycle	1
Bicycle	<1
Total private	**89**
PUBLIC	
Bus (service)	1
Train	2
Minibus	1
Coach (private/tour)	7
Total Public	**11**
Walked	1
Other	<1

Note: (1) figures do not add up to 100 %. due to rounding

Source: 1994 All Parks Visitor Survey: Lake District Site Survey

Study 2 The impact of visitors on the countryside

■ Seasonal unemployment

Tourism brings many benefits to the Lake District, but employment in the tourist industry is seasonal. The number of full-time jobs is fairly small compared to the size of the industry. Many young people gain summer employment working in hotels, bars and cafés, but by autumn most of these jobs are gone.

■ Footpath erosion

Road traffic is not the only environmental problem caused by tourists. Visitors spend 7 million visitor days each year walking in the fells for 4 hours or more. Walking has grown rapidly as a leisure activity, and many of the fells (or hillsides) in the Lake District are becoming trampled. Plants become trodden down and die, exposing bare soil, which in turn gets washed away by repeated rain showers. The result is that footpaths **erode** deeply (**Figure 3**), and become wider as people avoid the deep ruts.

Action can be taken to repair footpaths (**Figure 4**). The work is done by teams of volunteers and park wardens. It costs $4.2 million over a 10-year period to repair and maintain eroded paths. Well-maintained paths not only make for a more pleasant walking experience, but also protect wildlife habitats from disturbance as walkers use planned routes.

Figure 3 All those feet can wear out the footpaths ... **Figure 4** ... but they can be repaired – at a price.

Over to you

1. Summarise how, and why, there are traffic problems in the Lake District National Park.
2. **a** Use **Figure 2** to draw a graph to show how people get to the Lake District National Park.
   b Should more people be encouraged to use public transport i) to travel to the Lake District, ii) to get around the Lake District? Why?
3. Explain, in no more than 100 words, why tourism may bring economic problems.
4. Should action be taken to prevent so many homes being used for holiday lets or being bought as second homes? Explain your ideas.
5. In the USA, national parks have for years used wooden boarding to mark out footpaths and prevent erosion. What might be the advantages and disadvantages of doing this in the Lake District?
6. Is tourism in the Lake District worth all the problems that it brings?

RECREATION AND TOURISM

2.3 How should tourism develop in the future?

These two pages are about how tourism in the Lake District National Park should develop in future. They form a decision-making exercise, in which you are asked to consider issues from another person's viewpoint, and decide which is the best course of action.

In recent years, farmers in the Lake District National Park have suffered economically, as prices paid for farm produce have collapsed, followed early in 2001 by the worst outbreak of foot-and-mouth disease in two generations. In this exercise, you are a local planning officer who is asked to consider an application (**Figure 1**) from a farmer in the Lake District whose income has fallen drastically, and who is considering giving up altogether. Her bank have suggested that she has two valuable assets:

a a large barn which can be converted into some kind of accommodation, or housing;

b a field, close to the farm house and buildings, and into which it would be easy to lay power cables and water pipes. This would offer potential for some kind of tourist activity.

Her letter asking for advice from you, her local planning officer, is shown in **Figure 1**.

To help you, you have two sources. One, in **Figure 2**, gives different ideas for ways of using the barn and field. The second, in **Figure 3**, is a series of photos of different kinds of development.

```
                        Langdale View Farm
                              Langdale
                             Ambleside
                              Cumbria

Dear Sir or Madam

In recent years, I have been unable to
make much profit from my farm. The
foot-and-mouth outbreak in 2001 was
nearly the final blow for me. However,
I have a barn on my farm, which would
make good accommodation for tourists or
even a conference centre. I also have a
field close to the farmhouse and the
road, which would mean that I could get
electric cables, sewerage pipes, and
water supply on to my land. I could
then develop that as a campsite or
caravan park. Please advise me what you
think I should do, and which one you
think would get through the planning
office most easily.

Yours sincerely,

Mrs L. Jamieson
```

Figure 1 Mrs Jamieson's letter

Study 2 The impact of visitors on the countryside

Type of accommodation	Decide on a policy – should these be allowed or not?	Your reasons – consider economic and environmental effects
1 Using the barn		
New hotel or guesthouse		
Conversion of traditional barns into housing or business use		
Conversion of traditional barns into holiday accommodation		
Conversion of traditional barns into a bunkhouse for walkers, with basic facilities (running water, electricity, toilet)		
Youth hostel or outdoor pursuit centre for up to 20 people		
Conference centre for up to 20 people		
2 Using the field		
Static caravan sites – consisting of permanent caravans		
Construction of holiday parks consisting of chalet or A-frame buildings		
Touring-caravan parks (i.e. those hauled by cars)		
Tented camp sites (1) (i.e. large frame tents hauled by car)		
Tented camp sites (2) (i.e. small tents carried by walkers or cyclists)		

Figure 2 Judging different ideas for using Mrs Jamieson's barn and field

Figure 3 How do these different kinds of tourist accommodation affect the local environment and economy?

Over to you

1. As the local planning officer you have been given Mrs Jamieson's case to work on. Using **Figure 2**, consider the 6 possibilities given for using the barn, and the 5 for the field. Make a large copy of **Figure 2**, and decide for each one:
   a whether you would allow it, or not;
   b give reasons for your decisions.

2. Are there other possibilities you would consider for either the barn or the field?

3. Which single use would you allow for **a** the barn, **b** the field? Give your reasons in each case.

4. Write a letter back to Mrs Jamieson, outlining your decision about which use you think would be best, but also saying which other options you think the Lake District National Park would consider favourably.

RECREATION AND TOURISM

Study 3
Managing the countryside sustainably

This study looks at two areas – the Lake District National Park, and tourism in Zimbabwe. In the Lake District case study we will see how different people wanting different things can lead to conflict in the countryside and how this can be managed. In the Zimbabwe case study we consider ecotourism and see if this can help to provide more sustainable development in an LEDC.

3.1 Speed cameras on Windermere?

In March 2005, a ban preventing any boat users from exceeding 10 mph (15 kph) will come into force on all boats using Windermere (**Figures 1** and **2**), the largest lake in the Lake District National Park. Although a 10 mph speed limit was made law in 2000, there is a period of time for speedboat owners and businesses to adjust. Now that it is law, people are still debating whether or not there should be a ban.

The ban has caused howls of protest from some people, and their objections raise tempers among those who support it. It illustrates how different interests can lead to conflicts between different people. It depends upon who you are and what you want.

Windermere is famous for speed. In 1964, the late Sir Donald Campbell raced his famous speedboat 'Bluebird' on Windermere, and died when it crashed and overturned, before sinking. Each year, 'Records Week', a 4 km speed course, has been held on the lake, with its own exclusion zone to keep other craft out. Competitors are exempt from current noise and emission control byelaws. It draws in speed craft from all over the UK, and is valuable for local business. Some are keen to maintain it and do not want the ban.

No Exemptions From Speed Limit

There will be no exemptions from the speed limit for boats on Windermere when the new byelaw comes into force in four years' time. The organizers of motor boat races, speed trials and water-skiing events were told today that the 10mph speed limit would apply to all lake users. The only motorboats that will be allowed to break the speed limit are those involved in training for rescue boats and safety vessels.

It was recommended that an exemption could be made for Records Week. 600 letters of objection to the speed limit were received. But the needs of conservation, preservation of the peace of the countryside and the wishes of people who use the lake and the land around it led to the conclusion that no exemption should be granted. The event causes noise disturbance and excludes the public from large parts of the lake.

We are sorry that a small group of people will be affected by this decision, but we must think why this special area was designated a National Park in the first place, said Paul Tiplady, National Park Officer. 'Records Week' is noisy, and restricts other lake users from their legal right of navigation across Windermere.

Figure 1 Press release from the Lake District National Park, 2 January 2001

Figure 2 Boats using Windermere: should all kinds of boats and all speeds be allowed?

Study 3 Managing the countryside sustainably

a Supporters

- The Lake District National Park Authority
- Ambleside Civic Trust
- Ambleside and District Anglers Association
- Bowness Civic Trust
- Council for National Parks
- Countryside Agency
- Cumbria Wildlife Trust
- Friends of the Lake District
- Friends of Windermere
- Furness and South Cumbria Fisheries, Salmon and Trout Association
- South Windermere Sailing Club

b Opponents

- Holidays Afloat
- Shepherds Boatyard
- Waterhead Marine Ltd
- White Cross Bay Caravan Park
- Windermere and Bowness Chamber of Trade
- Windermere Lake Cruises
- Windermere Marina Village
- Windermere Water-Ski Association

Figure 3 Thumbs up or thumbs down? Groups supporting or opposing the speed ban

Figure 4 Two representatives put points for and against

a PAUL TIPLADY, National Park Officer

It is inevitable that some of the National Park Authority's decisions are controversial and we cannot please everybody. The Windermere speed limit is needed to ensure the safety of all lake users, and maintain the peace and quiet, which is valued by a majority of people. The letters of objection represent only a small percentage of power-boat owners on the lake.

b DAVID DALTON, lake user

We have a holiday home at Fall Barrow, our children have just started to water-ski, we live near York, North Yorkshire, we have a 1-year-old boat and my son has a dinghy; where can we go within a 2-hour drive? What about all the money we spend each week in Bowness-on-Windermere?

■ A speed ban – for or against?

The ban has its supporters (**Figure 3a**) who believe that Windermere is unsafe when powerboats are using the lake. Neither other boat or lake users, such as fishing enthusiasts, can enjoy their leisure as powerboats lead to exclusion from large areas of the lake. They believe that most visitors come to the Lake District for peace and quiet, and that powerboats ruin that. They claim that powerboats cause environmental damage, from the pollution of their exhausts, and aldo kill or harm wildlife, fish, and other freshwater species.

Meanwhile, the opponents of the ban (**Figure 3b**) believe that a voluntary code would be better and would result in the majority of users abiding by it. The Windermere Water-Ski Association claims that Windermere is an important training ground for champion British water-skiers such as Andy Mapple. To them, powerboats are as much a part of the 'cultural heritage' of the Lake District as any other activity. They feel that there are plenty of other lakes for more 'lower key' forms of recreation, and that Windermere represents a safe family environment for water-skiing. There are no alternatives to Windermere for speed on water – it is the longest lake. They believe that there should be room for all in a managed environment like a national park, and all users have a right of navigation, including those who wish to use powerboats.

Over to you

1. Using the information on these two pages write 200 words to summarize the case for, and against, a speed ban on Windermere.
2. Consider all the groups and people named in **Figures 3** and **4**.
   a. Decide whether their motives or concerns are social, economic or environmental.
   b. Draw a Venn diagram representing the three motives, 'economic,' 'environmental' and 'social'.
      - In the 'social' part, write any names of those whose motives are about people.
      - In the 'economic' part, write any names of those whose motives are about money or business.
      - In the 'environmental' part, write any names of those whose motives are about the environment, e.g. concerned with wildlife, or pollution.
      - Where you think people might overlap (e.g. social and environmental) write their names in the overlapping part.
   c. Is the speed ban mainly a social, economic or environmental issue?
3. Which arguments about the ban do you think are the strongest: **a** those in favour? **b** those against?
4. What is your view about the ban? Should it have been allowed? Justify your decision in 150 words.

RECREATION AND TOURISM

3.2 Can National Parks be protected?

One of the issues facing National Parks in the UK is that they are not wholly owned by one person, or one organization. Every national park in the UK is administered and managed by a national park authority, but it is owned by private landowners – farmers, hoteliers, private home owners. Each has their own view about how it ought to be run.

Compare this to the most used national parks in the USA, such as Grand Canyon or Yosemite. Each is owned and run on a non-profit basis by individual state governments, which can manage as they please. So, on approach to the Grand Canyon National Park, car owners are charged an entry fee, which goes towards some of the costs of managing the park in ways that are designed to protect the environment that people have come to see. The money pays for:

- car parks, which are screened out of view
- a park-and-ride transport system that prevents cars from access to the main park, and provides all-day transport around the park
- look-out points (**Figure 1**), from which the best views can be seen, and where visitor access can be made safe
- footpaths, which can be properly maintained without damage to either the natural environment or to people
- facilities such as toilets and washrooms, litter bins
- licences to refreshment providers (who charge a refundable deposit on drinks cans and bottles, thereby avoiding litter).

Figure 1 The Grand Canyon National Park, owned by the state government. How does this make a difference?

■ Managing conflict in the Lake District National Park

Many issues about National Parks raise all kinds of feelings. Some people object to most developments in the Lake District because they believe that the environment is special and should not be changed. Others feel that development is important for people who already live there and for whom economic development means jobs. The job of the Lake District National Park Authority is to try to manage these views, because they **conflict** – that is, they are opposed.

Land owners might want to take advantage of tourism, such as providing a caravan site. Others might object to this. Alternatively, farmers might find that tourists cause damage, such as leaving gates open or trampling on crops. Tourists might object to farmers' proposals to stop them walking on public footpaths. The National Park Authority helps to deal with these conflicts, but it has little power over them. All it can do is try to resolve problems, and prevent any development which it feels is not in keeping with the landscape. It is a kind of guardian.

Seeing the National Park through different eyes:

Figure 2 ...landscapes at Coniston Water ...

Study 3 Managing the countryside sustainably

Figure 3 ... farming Beckside Farm ...

Figure 4 ... and hiking at Langdale

Study the photographs in **Figures 2** to **4**. For each one, think about:

a what the landscape is for   **b** who it is for.

It should be possible very quickly to see that some of your views about what it is for, and for whom, may conflict.

For instance, **Figure 2** shows Coniston, one of the most popular lakes. You have already seen how a speedboat ban on Windermere has led to conflict. Consider what other uses of Coniston might cause conflict among local people or tourists. Then consider the sheep farmer in **Figure 3**. What difficulties might the farmer face on this land as a result of tourism? Perhaps there are other more profitable activities that this land could be used for. Think about **Figure 4**. What is it about the landscape that these walkers have come to see? What threats are there to this landscape? Imagine how the view would change if a water company built a dam for a reservoir. Or if a quarrying company wanted stone from the valley side.

It is possible to show how different people feel about an issue by plotting a **conflict matrix** like that shown in **Figure 5**. To draw up a conflict matrix, follow the instructions on page 73.

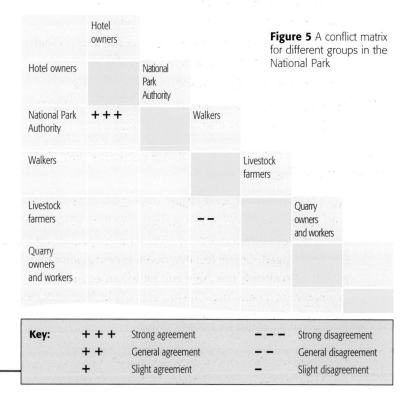

Figure 5 A conflict matrix for different groups in the National Park

Key:	+++	Strong agreement	---	Strong disagreement
	++	General agreement	--	General disagreement
	+	Slight agreement	-	Slight disagreement

Over to you

1. Make a blank copy of the conflict matrix in **Figure 5** in your book. On it:
   - think of any of the people or groups that you have studied, and who would have an opinion about *what* the Lake District is for, and *who* it is for
   - write their names into the left-hand column of the matrix, and again at the top of a column on the right
   - decide who agrees with who, and how much, by using + or − symbols as shown in the key.

2. In a few sentences, explain what the table is about and how you drew it up.

3. Describe what the table shows about:
   a who agrees, and
   b who disagrees with how the Lake District should be run.

4. Select any two people who strongly agree, and two who strongly disagree. Suggest why these people agree or disagree with each other.

RECREATION AND TOURISM

3.3 Tourism in Zimbabwe – 1

■ Helping rural development and conservation

Ecotourism is one of the world's popular concepts in tourism. It means a kind of tourism that does not impact badly on the environment. Instead, it uses the environment as an attraction, and seeks to preserve it. Sections 3.3 and 3.4 are about a project in Zimbabwe, an LEDC, where tourism is being used as a means of helping economic development to take place. It uses ideas about what is '**sustainable development**' and what is not. Sustainable development means being able to improve the economy and jobs but not in ways that cause environmental damage, or ways that do not last.

Zimbabwe is one of the world's great tourist attractions, though so far it has avoided mass tourism. On offer are white-water rafting along the rapids of the Zambezi, scenery such as the Victoria Falls, safaris to see leopards or elephants, or over 600 species of birds. Its natural vegetation of savanna woodland, a mix of tall grasses, shrubs and trees (**Figure 1**) provides excellent wildlife habitats. The country is well positioned to take advantage of nature-based tourism.

But Zimbabwe is also one of the world's poorest countries. The poorest land with least farming potential (42% of the country) supports nearly half of the country's population. These areas receive low rainfall, (450–600 mm) and have poor soils. Crop production in these regions is very low and the areas are really only suited to cattle and wildlife ranching. Shallow, poor soils and steep slopes make these areas liable to soil erosion. But it is in these lands that the greatest tourist potential exists - most of Zimbabwe's national parks are found in these drier areas (**Figure 3**).

CAMPFIRE is a rural development project in Zimbabwe which aims to put control of tourism and conservation into the hands of local people. Under Zimbabwe law, property holders can claim ownership of wildlife on their land and manage it so that they benefit from its use. It aims to:

- improve standards of living, by helping the local economy
- provide an incentive for **conservation**.

■ How do communities benefit from tourism?

Usually, local communities grant permits to tour companies in return for payment. A local tour operator organizes photographic safaris in the area for small groups of tourists, as well as cultural visits to traditional villages, where local people sell handicrafts and souvenirs.

Figure 1 Two of the 'locals' in the Savanna in Zimbabwe

Zimbabwe Factfile

Population – 11.4 million people
Population increase – each year 2.7 %
Life expectancy – 51 years
Income (GNP) per person – US$540 in 1995
Literacy – 90% of men and 80% of women can read and write
Largely rural – only 31% live in cities

Figure 2 Zimbabwe factfile

Figure 3 Zimbabwe and its national parks

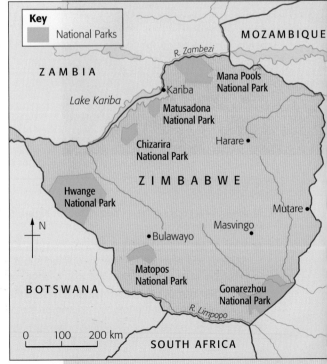

Study 3 Managing the countryside

Figure 4 Hunting the best snaps: a photographic safari

A few communities host tourists themselves. Sunungukai Camp, a few hours from Harare, is managed by local people. As well as receiving income, locals work as guides, taking people to enjoy the view, or mountain hiking, catch fish, or to watch birds, hippos, crocodiles, and small game, or to see local bush paintings. Others make handicrafts and souvenirs to sell at the camp. Tourists experience local culture by sharing food, music and lifestyle. The more adventurous stay in mud huts, eat local dishes, and try tasks such as grinding millet, ploughing, and mat-making.

There are other ways in which local people make money from wildlife on their land, such as granting licences for elephant hunting. This will be explored in section 3.4. Other ways include:

- 'Harvesting' resources such as crocodile eggs, timber, river-sand and caterpillars.
- Live animal sales to commercial game reserves or national parks. One council sold ten roan antelope, earning US$50 000.
- Meat sales from wildlife, sold to neighbouring communities or towns.

■ What is the impact of tourism upon Binga?

People in Binga district, a rural area in north-western Zimbabwe, live in chronic poverty. Before the CAMPFIRE project, they rarely had enough grain, existing on wild foods, drought relief, payments from relatives working in towns, and selling any remaining crops and livestock. When wildlife damaged their crops or livestock, it ruined their livelihoods. Now, the CAMPFIRE project has given them money to build solar-powered electric fences around agricultural land and villages to keep out wildlife. Wild animals are now vital to the area: income from fishing, hunting and tourism tops up household earnings. One estimate shows that CAMPFIRE has increased incomes by up to 25%.

Income from the project goes to households, who decide how to use the money. Before CAMPFIRE, the Binga district had only 13 primary schools and no secondary schools. Now it has 56 primary schools and nine secondary schools. CAMPFIRE income has also been used for: drilling wells for clean water; grinding mills for local grain; building health clinics; fencing arable and residential land.

Over to you

1. Select four features that you think make tourism in Zimbabwe unique. Using this information, design a tourist poster to be displaced throughout a major urban transport system in the UK, such as the Manchester Metro or London Underground.
2. Summarize the aims and methods used in establishing the CAMPFIRE project, and where it is taking place.

Include these words (in any order): tourism, rural development, soil erosion, national parks, permits for tour companies, wildlife, local economy, conservation, licences, local culture.

3. Draw up a table of the CAMPFIRE project showing its advantages and disadvantages. Which are greater?

RECREATION AND TOURISM

3.4 Tourism in Zimbabwe – 2

■ African elephant – friend or foe?

One of central Africa's images is that of a wildlife haven. Few animals illustrate this better than the elephant. But in Zimbabwe, the elephant has a varied reputation. Almost 10 000 of them roam Zimbabwe's dry lands, often wandering from national parks into farmland, eating or trampling crops, and destroying families' income or food. Elephants also threaten people's lives – elephants or buffaloes in Kariba have killed over 100 people since 1980.

But globally, **conservation** of wildlife is a popular and important issue and Zimbabwe has an excellent conservation record. In 1900 its elephant population was less than 4000. Now it is over 64 000 in total, twice the amount that it can sustainably support without severe environmental destruction, and the loss of plant species. Now, some control is necessary. How can this be done in a way that supports ecotourism?

■ Elephants and the CAMPFIRE project

Before the CAMPFIRE project, rural Zimbabweans feared elephants, and saw little reason to conserve them. Now, elephants are seen as friends – they are tourist attractions and a source of income from hunting. Over 90% of CAMPFIRE income comes from foreign hunters who come to Zimbabwe to hunt elephants, buffaloes, lions or other wild animals. This particular kind of hunting is known as **trophy hunting**.

- Trophy hunting is considered the 'ultimate' form of ecotourism in Zimbabwe. Rural communities earn good money from foreign hunters, who pay high fees to hunt. Hunters travel in smaller numbers, and are usually satisfied with more basic amenities than other tourists, so they have a less damaging impact on the environment.
- Licensed hunting is profitable for rural communities. To them, trophy hunting is better than photographic tourism, as it provides much higher income.
- Hunters help to prevent poaching, because hunting permits are only granted to professional safari hunting companies.
- The poorest lands have the majority of the elephant trophy hunting market, so it benefits Zimbabwe's poorest the most.

In some districts, elephant hunting is the only source of income. Each elephant killed brings a trophy fee of US $12 000, and daily hunting fees of $1000. A single elephant can bring in $33 000 on a 21-day hunt. The benefits for communities from the income can include new schools, a new grinding mill, and a Z$200 cash payment to each household. This is a large income – a rural family of eight people would normally get by on about Z$150 per year in Zimbabwe.

Figure 1 This is what elephants can do. How would you feel if these were *your* crops

Figure 2 Game hunting: not popular with some, but what are the benefits?

Study 3 Managing the countryside sustainably

> *Ten years ago, we liked the animals. But now we like them more because we are getting money for them.*
> Arius Chipere, Dete village wildlife committee

> *When we are hungry, the elephant is food: when we're full, the elephant is beautiful.*
> Elderly villager, Zambezi Valley

> *Before CAMPFIRE, we thought we had too many animals, always stealing our food. We did not count them. Now we are happy with almost every elephant we see. Our neighbours complain that they do not have enough!*
> Phineas Chauke, Tongagara

Figure 3 Changing views about elephants

Figure 4 The alternative to regulated hunting: the results of elephant poaching

What effect has CAMPFIRE had on wildlife?

The CAMPFIRE project has helped elephant conservation.

- Rural people have changed their views about elephants on their land (**Figure 3**). In one district, the number of elephants killed for damaging crops dropped from 45 to 3 in only three years.
- In another area, villagers have dug waterholes and arranged food for elephants in times of drought.
- Communities are trying out wildlife management techniques, to prevent habitats being destroyed by farming, and uncontrolled hunting or poaching. CAMPFIRE issue hunting licences on the basis of strict quotas each year. Estimates of wildebeest, sable, elephant, lion, and leopard populations are made from surveys, and from hunters. Villagers are encouraged to map how many animals are seen, so that hunting quotas can be set.

Culling elephants - what do you think?

Killing elephants is a sensitive issue in other countries. In the 1980s, countries such as the UK banned imports of ivory and elephant products because they thought it would stop elephant poaching. However, bans on the import of rhino horn have not stopped poaching, whereas local people involved in sustainable management of wildlife may do so.

The ban on elephant products has badly affected rural communities in Zimbabwe. One district could have quadrupled its income if it had been allowed to sell ivory and hides from three elephants hunted after raiding crops. Rural communities have US $1.6 million in stockpiled ivory in Zimbabwe. If they cannot earn money from elephant products, poor families will clear wildlife habitats to make way for more croplands. Zimbabweans feel that:

- elephants are not endangered in Southern Africa
- income from elephant management helps to support the economy
- trophy hunting is seen in Southern Africa as a form of ecotourism
- the realities of living with elephants is not the image seen on TV
- the ivory ban does not help elephant conservation.

Over to you

1 'Sustainable development' means being able to improve the economy and jobs but not in ways that cause environmental damage.
   a In your own words, say whether you think the CAMPFIRE project:
      i) improves the economy and jobs
      ii) reduces environmental damage for the future.
   b Define the following words, and say what they have to do with 'sustainable' policies about wildlife: quotas, licences, environmental destruction, habitats, ecotourism.
2 The CAMPFIRE project might conflict with your attitudes towards African animals.
   a Conduct a survey in your year group, and at home, to find out how people feel about trophy hunting, 'harvesting' resources such as crocodile eggs, selling live animals to commercial game reserves or national parks, and meat sales from antelope or zebra.
   b What do the results tell you?
3 After your survey, write 150 words on each of the following:
   a how I feel about hunting animals such as elephants;
   b how much I agree with what the CAMPFIRE project is doing for rural communities;
   c whether it is right that I or anyone else in the UK should protest about animal hunting in Zimbabwe.

OS MAP SYMBOLS

ROADS AND PATHS

Not necessarily rights of way

- Service area M 8, Elevated, Junction number — Motorway (dual carriageway)
- Motorway under construction
- Unfenced, Footbridge, A 92 (T) — Trunk road
- Dual carriageway, A 917 — Main road
- Main road under construction
- B 9131 — Secondary road
- A 855, B 885 — Narrow road with passing places
- Bridge — Road generally more than 4 m wide
- Road generally less than 4 m wide
- Other road, drive or track
- Path
- Gates, Road tunnel
- Ferry P — Ferry (passenger)
- Ferry V — Ferry (vehicle)

PUBLIC RIGHTS OF WAY

(Not applicable to Scotland)

- Footpath
- Bridleway
- Road used as a public footpath
- Byway open to all traffic

RAILWAYS

- Track multiple or single
- Track narrow gauge
- Bridges, Footbridge
- Tunnel
- Viaduct
- Freight line, siding or tramway
- Station (a) principal (b) closed to passengers
- LC Level crossing
- Embankment
- Cutting

BOUNDARIES

- National
- London Borough
- National Park or Forest Park
- NT National Trust
- County, Region or Islands Area
- District
- NT always open
- NT limited access, observe local signs

GENERAL FEATURES

- Electricity transmission line (with pylons spaced conventionally)
- Pipe line (arrow indicates direction of flow)
- ruin Buildings
- Public buildings (selected)
- Bus or coach station
- Coniferous wood
- Non-coniferous wood
- Mixed wood
- Orchard
- Park or ornamental grounds
- Quarry
- Spoil heap, refuse tip or dump
- Radio or TV mast
- Places of Worship — with tower / with spire, minaret or dome / without such additions
- Chimney or tower
- Glasshouse
- Graticule intersection at 5' intervals
- Heliport
- Triangulation pillar
- Windmill with or without sails
- Windpump/wind generator

HEIGHTS/ROCK FEATURES

- 50 — Contours are at 10 metres vertical interval
- ·144 — Heights are to the nearest metre above mean sea level
- outcrop, cliff, scree

WATER FEATURES

Marsh or salting, Towpath, Lock, Slopes, Cliff, High water mark, Low water mark, Aqueduct, Canal, Ford, Flat rock, Lighthouse (in use), Weir, Normal tidal limit, Sand, Beacon, Lake, Bridge, Footbridge, Dunes, Lighthouse (disused), Shingle, Mud
- Canal (dry)

ABREVIATIONS

P	Post office	PC	Public convenience (in rural areas)
PH	Public house	TH	Town Hall, Guildhall (or equivalent)
MS	Milestone	Sch	School
MP	Milepost	Coll	College
CH	Clubhouse	Mus	Museum
CG	Coastguard	Cemy	Cemetery
		Fm	Farm

ANTIQUITIES

- VILLA Roman
- Castle Non-Roman
- Battlefield (with date)
- Tumulus

TOURIST INFORMATION

- Information centre, all year/seasonal
- Viewpoint
- Parking
- Picnic site
- Camp site
- Caravan site
- Youth hostel
- Telephone, public/motoring organisation
- Golf course or links
- PC Public convenience (in rural areas)

GLOSSARY

Abrasion — Mechanical wearing away or erosion caused by rubbing and scouring effect of material carried by rivers, glaciers, waves and the wind

Accessible — A place or feature to which people can get easily by some form of transport or by walking

Acid rain — Rainwater containing chemicals that result from the burning of fossil fuels

Affordable homes — Homes which are provided at below the market price. Developers are now under obligation to build a certain number on large estates. Housing Associations also offer homes on a part purchase and part rent basis for people unable to afford mortgages

Ageing population — The average age of the people in this country is getting older. There are more old people living longer. There may also be fewer babies being born.

Agglomeration economies — When industries benefit from locating close to each other

Agri-business — Commercial farming operations that try to maximize profit from farm land by using machinery and chemical sprays

Agro-chemical — Chemicals produced for the farming industry

Aquifer — Layer of rock which can store and transmit water

Aquifer recharge — Replacement by rainfall of water which has been abstracted from the aquifer

Arable — Crop growing

Artificial fertilizer — Fertilizers produced from manufactured chemicals (instead of manure)

Asylum seekers — People who have come to a new country and ask to stay because they are afraid that they would be killed or persecuted if they returned home

Beach nourishment — Any process that helps to encourage the growth in size of a beach

Biodiversity — The range of plants and animals living in an area

Birth rate — The number of births per thousand in a population per year

Borehole — A drilled hole from which water can be obtained

Brandt Report on World Development — A report published in 1980 that defined a world view of 'rich' north and 'poor' south

Brownfield site — A site which has been used for buildings or other development and has been left to run-down/become derelict. It will need to be improved or cleared before it can be used again

Car pooling — The sharing of car transport by several commuters

CBD — Central Business District. The commercial and business centre of a town or city

Cliff face — The side of the cliff between the top and foot

Cliff foot — The base of a cliff, which is normally attacked by waves

Climate — The average of weather conditions over a period of time, at least 30 years

Cohort — A group of people of the same ages. Usually in 5 year age groups, e.g. 10–14, 15–19

Collision margins — Found where plates of continental crust move towards each other. The edges of the plates buckle upwards to form large mountainous areas

Commute — Travelling a considerable distance to one's workplace

Conflict — Where two or more people or organisations disagree about a proposal

Conflict matrix — A diagram designed to show the strength of feeling about a development or proposal, and who agrees or disagrees with it

Conservation — Policies that protect physical or human features from further change, e.g. by protecting buildings or wildlife

Conservative margins — Found where two plates move alongside each other, crust is neither formed nor destroyed

Conserve — To use carefully so that it lasts longer. To manage so that it does not run out

Constructive margins — Found where plates move away from each other and new crust is formed

GLOSSARY

Conurbation A city which has grown so much that it absorbs other towns to form a continuous urban area

Corrasion The same as abrasion

Death rate The number of deaths per thousand in a population per year

Demand management Managing water resources by reducing the amount of water used by technology, economic or financial means

Demographic Transition Model A model which shows changes in birth rate, death rate and population size over a period of time

Dependents The people who have no job and no income and rely on others to provide for them, e.g. the young and the elderly

Dependency ratio The proportion of young and elderly (dependents) to the economically active (workers) in a country. The dependents rely on the workers to provide money for all they need – education, healthcare, pension. The government in the UK is trying to persuade people to provide medical insurance and private pensions for themselves when they are old so that they do not need so much support.

Desertification The process by which desert regions spread to surrounding areas, e.g. the Sahel in west Africa

Designer Outlet Shopping development where goods are retailed at less than the normal price

Destructive margins Found where plates of oceanic crust collide with plates of continental crust. The oceanic crust is forced downwards and is destroyed

Diguette A line of stones used in projects in Africa to trap water running off the land in rain showers. In damming the water to a height of a few centimetres, soil particles are trapped and settle, which helps to reduce soil erosion

Distance decay Where something reduces the further you get from a place

Ecology The relationship of plants and animals with each other and the natural environment

Economically active population The people who have jobs and are adding to the wealth of a place

Ecosystem A community of plants and animals together with the environment in which they live

Ecotourism A form of tourism that is designed to reduce harm to the environment, e.g. by attracting people to see its special qualities without harming what they come to see

Environmental impact The effect that something has on the land, air, water, scenery, soils

Environmentally friendly Not harmful to the land, air, water, soils, scenery

Epicentre The point on the earth's surface immediately above the focus of the earthquake

Erosion The wearing away of land by water, ice or wind

Evaporation The transfer of water to the air as water vapour

Exponential A mathematical word that means something increases by doubling, i.e. 2,4,8,16

Farming system A way of studying farms, that includes inputs (what is put into the land) and outputs (what is produced by the land)

Favela An informal squatter settlement in Brazil

Fetch The length or stretch of water between two land areas, which is one of the factors which decides wave height and strength

Flood plain The wide flat valley floor of a river which the river flows over in times of flood

Focus The point where seismic waves start and the rocks give way

Fold mountains High mountain ranges formed when two plates move together

Footloose industries Industries that are not fixed to locating in a particular area. This could be because they are not tied to raw materials for example

Formal economy Legitimate businesses where companies or organisations trade legally and are registered with the government

Glossary

Formal employment Employment often controlled by businesses in the formal economy offering permanent jobs

Full-time work People working over 36 hours per week and who are paid either weekly or monthly. Hours are usually fixed and workers benefit from paid holidays and paid sick leave

Gabions Smaller boulders contained inside metal cages, placed at the foot of a cliff or along a beach to protect it from further erosion

Genetically modified A process by which scientists try to transfer the qualities in one species into another, using gene transfer. In this way, genes can be transferred between species which could never breed in nature in order to try and produce 'super-crops'. Note that this is not the same as 'hybrid'

Geology The study of rocks, or the rock type of an area

Glaciers Moving flows of ice found in mountainous areas or near to the north or south poles (e.g. on Greenland or Antarctica)

Global warming The increase in the world's temperature thought to result from the release of carbon dioxide and other gases into the atmosphere by the burning of fossil fuels

Green corridor Area of open land on either side of a river. It has been left undeveloped because it is liable to flooding

Greenfield site Land which has not been built on but which has been designated for development

Greenhouse gases e.g Carbon dioxide, methane, CFCs, nitrogen oxides. These are the major contributors to global warming

Gross Domestic Product (GDP) The value of all the goods and services sold by a country in one year

Groundwater Water stored beneath the soil in aquifers

Growth rate The increase or decrease in population number per thousand per year

Groynes Wooden or stone structures on a beach, built at right angles to the shore in order to trap sand and shingle moving by longshore drift

Habitat The place in which a plant or animal is suited to living

Hard engineering Anything that is built to protect a cliff, or coastline, such as a sea wall or revetment

Herbicide Chemical sprays used to kill unwanted plants in crop fields

Honey-pot A place that attracts tourists in large numbers

Hotspots Popular tourist areas. Climate may be a major factor in their development, but other factors such as scenery, culture and good facilities are also important

Hybrid A cross-breed of plant or animal. Note that this is not the same as 'genetically-modified' species

Hydraulic pressure The compression of air inside cracks and fractures when a wave breaks against a cliff. As the waves retreat, they release the air with an explosive effect that helps to break down and shatter rock over time

Hydrograph A graph showing changes in the discharge of a river over time

Impermeable A term used to describe rocks whose particles are so compressed together that they do not allow water through

Infiltration The movement of water into soil surface and underlying rocks

Informal economy Businesses that operate outside legal restrictions and are unknown to (or at least not tracked by) tax or health and safety officers

Informal employment Employment in the informal economy which depends on individual's initiative, e.g. jobs such as shoe cleaning, selling things on the streets.

Intensive farming A system of farming that uses large quantities of fertilizer and machinery in order to increase crop yields or output from a farm

Interception When it rains water droplets are stopped from reaching the ground by trees and plants. Interception is greatest in the summer

GLOSSARY

Irrigation   Watering the land by means of channels, sprinklers and drip systems

Landfill site   A large depression in the ground which is used for dumping waste and which at the same time is being reclaimed for future development

Less developed   Many people work in primary employment, low levels of technology

Life expectancy   The average age that a person might live to. People live to be older in MEDCs than in LEDCs

Local Agenda 21   An international agreement put in place at a small-scale level such as a town. The agreement is to manage the environment in a more sustainable way, so it is damaged as little as possible

Longshore drift   The movement of beach material along a coast, caused when waves break onto the shore at an angle. Beach material is then transported along the direction of wave movement

Malnutrition   Eating a diet which is unbalanced and likely to lead to ill-health

Managed retreat   A policy for managing coasts where the decision is to reduce the amount of coastal protection. In some cases, protection may be very limited and low cost; in others it may be withdrawn altogether

Mass movement   Downhill movement of weathered material, such as landslides, slumping or rock falls

Mechanisation   The process by which machines have taken over jobs formerly done by people

Microclimate   The local climate of a small area caused by, for example, hills and valleys generating temperature, wind and precipitation conditions that may differ from the general pattern

Migrant   A person who moves from one country to another to live

Migrate   To move; e.g. those who move from rural to urban areas to live

Migration   The movement of people from one place to another

Model (e.g. Demographic Transition)   A simplified way of considering a complicated problem. Often as a diagram or an equation

More developed   Fewer people work in primary employment, more wealth, more technology, good communications

National Park   An area which is protected from further development because it is considered special. In some countries (e.g. USA, Australia) these areas are bought and managed by the government; in others (e.g. the UK) they are owned by private landowners and protected by local authorities

NIMBYism   An attitude of not wanting development in one's own area (Not In My Back Yard) but it can take place elsewhere

Nomadic   People who wander in search of a living

Non-renewable   Not able to be replaced in the time that it will be used up. It will run out. Another word for it would be 'finite' e.g. oil

Organic farming   Methods of farming that do not involve any artificial fertilizers, herbicides or pesticides

Part-time work   People working for, perhaps, one or two days a week, or for a few hours every day. Part-time workers receive hourly rates of pay rather than salaries, and are paid according to the number of hours they work. Many get few benefits, such as paid holidays, and may not receive sick pay

Permeable   The ability of a rock to allow water through along cracks and fractures

Pesticide   Chemical sprays used to kill insects which would otherwise feed on crops

Plate tectonics   The theory that the earth's surface is divided into plates consisting of continental and oceanic crust

Plates   Segments of the earth's crust

Population   The people living in a place

Population processes   The actions that change the number of people in a population: births, deaths, migration

Glossary

Population structure The number of people in each age group. It may also be the number of males and females. Often shown by a population pyramid

Primary employment Jobs based around the collection, or harvesting of raw materials, e.g. farmer, miner, and lumberjack

Pull factors The things which draw people to the places they migrate to

Push factors The things which are wrong and which encourage people to leave their homes and migrate

Pyroclastic flow River of hot gas, ash, mud and rock moving at very high speeds at temperatures of about 500 °C

Quality of life A measure of people's well-being and of what their life is like

Quaternary employment Jobs which provide information and expertise, e.g. market researcher and IT consultant

Quota A limit on production. The European Union sets limits on the amount of milk that can be produced by EU farmers. Farmers who exceed these are penalised financially

Rate of natural increase The change in the total number of people per thousand per year. It may be positive or negative, an increase or a decrease

Recycle Give some items back so that they can be used again and do not use up resources so quickly. E.g. glass can be melted down and made into new glass. Old newspaper can be made into egg boxes.

Refugees People who have left their home country and have no home. They may leave because of war, famine, hazard or political problems

Renewable Can be used again and again as it can be replaced more quickly than it is used up, e.g. trees for wood

Renewable energy Energy sources that do not run out, such as those that come from sun, wind, or energy from water

Revetments Large boulders placed at the foot of a cliff, used to protect a cliff against further erosion by waves

Richter Scale A scale from 0–8 used for measuring the magnitude of earthquakes based on recordings from a seismograph 100 km from the epicentre of the earthquake. The larger the number the bigger the earthquake

Runoff Water flowing over the surface of the land

Sahel An area of western Africa on the edge of the Sahara Desert which has suffered climate change and drought in recent decades, so that the desert has spread

Salt marsh An area of low-lying mud and sand within an estuary, usually formed behind a spit on the sheltered side of a bay

Secondary employment Jobs involved in manufacturing, or making things, e.g. car assembly, carpenters, builders, and potters

Sedimentary rock A rock which has been formed from fragments of broken down igneous rocks. The fragments have been deposited in layers on land or in water

Set-aside A grant paid to farmers by the European Union in return for not planting their land. This is used when certain crops are being over-produced within the EU

Site of Special Scientific Interest (SSI) An area with special or unique environmental qualities that is considered worthy of protection, e.g. containing rare plants or animals

Soft engineering Anything that protects a cliff or coastline using natural processes, such as the development of a beach through beach nourishment

Soil erosion The removal of soil from the land surface by wind or water

Spit A landform formed from sand and shingle at the end of a beach, often across an estuary or bay. It is created by longshore drift

Stable (population) A population which changes very little in total size. It has the same number of people all of the time

Standard of living The level at which people exist each day. A high standard of living is plenty of food, clean water, machinery to do difficult tasks, extra money for holidays. A low standard of living is poor housing, dirty water, no sewerage, hand work, poverty

GLOSSARY

Stewardship Looking after the environment. Managing it well so that it can be used and enjoyed by the next generation

Strata Layers of rock

Sub-aerial processes A term used to describe all the processes that affect a cliff face, such as weathering or movement of the cliff (e.g. slumping), or erosion by rainwater

Subduction zone Found at a destructive plate margin where oceanic crust moving towards continental crust is forced downwards into the mantle and destroyed

Subsistence A method of farming in which only sufficient food is produced to feed those who farm it; there is no surplus

Substitution Providing a different resource/an alternative

Sun-lust tourism Tourism affected by climate, sometimes called '3S' tourism by the travel trade – sun, sand and sea

Sustainable A way of using resources to ensure there is sufficient to be used in the future

Sustainable development Developments which meet the needs of today but which have minimal impact on both people and environment. The needs of future generations must not be impaired

Tertiary employment Jobs involved in providing a service, e.g. teacher, librarian, financial adviser, waiter, or supermarket worker

Throughflow The movement of water through soil and rock

Topsoil The upper layer of soil which is most fertile

Tourism An industry that is based on people's wish to travel and spend leisure time away from home

Transnational corporations (also known as multi-national companies) Large companies that operate in more than one country

Transport corridor Routeway where there is both road and rail transport, and possibly water transport

Trophy hunting A form of hunting that seeks out large animals such as elephant, carried out using a system of permits or licences

Urban fringe Edge of the built up town/city

Urban heat island A zone of higher air temperatures found in and around urban areas

Urban renewal Transformation of decayed urban centres by new flagship developments to improve the environment and stimulate new growth (economic and cultural)

Urbanization the increase in the percentage of the population living in towns and cities

Virtual water Water which is transferred by global trading of crops which require large amounts of water for growth. By importing for example wheat grown in regions of water surplus, water short countries are virtually importing water but more easily and cheaply than by transporting massive volumes of water. (It needs 1000 cubic metres of water to produce one tonne of wheat)

Volunteer A person who offers to do a task/job without being paid

Water cycle The continuous circulation of water between the sea, atmosphere and land stores

Water resource A supply of water

Water stress Insufficient water to meet the requirements of a particular country or region

Water table The upper level of water stored in an aquifer

Weather The short term day-to-day state of the atmosphere

Weathering The gradual break down of rocks at the earth's surface by physical means (such as frost) or chemical (such as the effects of acid rain)

World Heritage Site (WHS) A site or area recognised by the United Nations as having special qualities, and which ought to be protected from further development

INDEX

A

acid rain 129, 182, 186–189, 190
    China 187
    Europe 187–189
    Black Forest, Germany 188
    India 187
    Norway 189
    UK 189, 193–194
acid soil 188
Africa (irrigation) 183
agri-businesses 201–202, 208, 214
agriculture (see farming)
agro-chemicals 212
AIDS 111
Alton Towers, UK (tourism) 225
Andes Fold Mountains (volcanoes and earthquakes) 80
Ambleside, UK (tourism) 231, 234
animal welfare 201
aqueducts 160
aquifers 150, 172
Aral Sea (water management) 170–171
Aral Vision 171
Ashford, UK (settlement) 8–13
Aswan Dam, Egypt 162–163
Aswan Dam (High), Egypt 157, 163–165
    advantages 164
    disadvantages 165
Aswan, Egypt (water supply) 152
    (water management) 162–165
asylum seekers 114, 117
Atlantic Sea (earth's plates) 80
Australia (tourism) 226
Aviaries Farm, UK (organic farm) 206–207

B

Baia Mare gold-mine reservoir (pollution) 168
Bangkok (global warming) 192
Bangladesh (Ford) 37
    (flooding) 99, 101
    (global warming) 193
Barrier Reef (coastal protection) 66, 69, 76
Barton-On-Sea, UK (coastal erosion) 54–65
beach nourishment 59, 62, 64
beach protection 60–65, 69
Beijing, China (water demand) 172
Belgrade (water supply) 168
bid-rent theory 49
Binga, Zimbabwe (impact of tourism) 243
biodiversity 68, 76, 205, 209
Birmingham, UK (transport system) 24–25
Black Forest, Germany (acid rain) 188
Boserup, Ester 140
Bosnia (migration) 117
Boston, USA (transport system) 26–27
brownfield sites 6, 8, 10, 12–13, 145
BSE 199, 208
Burkina Faso, Africa (farming) 214–215, 218–219

C

Cairngorms (weather) 177
CAMPFIRE project 242–245
Canada (tourism) 226
Caribbean (tourism) 176
Central Arizona Project (CAP) 161
Chad (drought) 217
Cherrapunji, India (water supply) 152
China (acid rain) 187
    (global warming) 172, 194
    (hydro-electric power) 178
    (population) 102, 104–105, 146–147, 158–159
Christchurch Bay (coastal protection) 54–66
Circle of Development 119–120
cities 14, 22–23
    climate 184–185
    sustainability 7, 22
    traffic 24–27
CJD 199
classification 126
Climate Change Impact Group 193
climate 152, 163, 171–197, 230
    change 190–196, 217
    modification 182
    as resource 196
    sustainability 190, 194–195
    tundra 194
Club of Rome 140
coasts 54–77, 192, 195
    Christchurch Bay, UK 54–66
    Daintree, Australia 66–77
Coca Cola (TNC) 35
commuting 7, 23–24
conflict matrix 73, 241
conservation 76, 98, 187, 190, 196, 242–245
continental crust 80–1
conurbation 17
convection currents 79
Cornwall (tourism) 224–225
countryside 6–7, 11–12, 222–245
    management 226, 238
    sustainability 6, 238
CPRA (Council for the Preservation of Rural England) 6
Crow Report 9
Cyclone Eline 98

D

Daintree coast, Australia
    (coastal protection) 66–77
    (sustainable development) 74, 76–7
Dakupa project 220
dams 100, 162–165, 173, 230
    advantages 164
    Aswan Dam, Egypt 162–163
    Aswan Dam, Egypt (High) 157, 163–165
    disadvantages 165
    Hoover Dam, USA 161
    Three Gorges, China 178
deforestation 195, 220
demographic transition model 111
Denmark (wind power) 179
deposition, dry 186
Derince, Turkey (earthquake) 86
development, sustainable 22, 74, 76–77, 221, 236–245, 242
Dhaka (global warming) 192
diguettes 220–221
distance decay 225
Donana, Spain (water management) 173
drought 15, 153, 174, 178, 180, 182, 216–217, 220
    Chad, Africa 217
    Eritrea, Africa 216
    Mauritania, Africa 217
    New Zealand 178
    Niger, Africa 217
    Sahel, Africa 216–7
    UK 153
Dubai (tourism) 176
Dzaleka refugee camp 115

E

earth's mantle 79–81
earthquakes 78–81, 86–89, 117
    Derince, Turkey 86
    effects 86–87, 89
    epicentre 86
    Gujarat State, India 78
    patterns 78–79
    planning 89
    prediction 88–89
    Washington State, USA 78
ecology 62, 68–69, 226
economy 65, 67, 180
    agglomeration 53
    formal 31
    'grey pound' 223
    informal 31
    service-based 32, 42
ecosystems 69, 180, 193
ecotourism 242–5
Egypt (population) 156–157,
    (water supply) 162–165
emigration 6
emissions 179, 187, 189
employment 8, 13, 15, 18–21, 23, 30–53, 70, 72, 119
    primary 30, 32
    quaternary 30
    Reading, UK 42–53
    secondary 30
    South East Asia 35, 37–41
    South Korea 32, 38
    tertiary 30
    UK 30–31
    women 36
energy 128, 139, 144, 149, 155, 195
    alternative sources 178, 195
    biogas plants 132–133, 136–137
    consumption 138, 149
    fossil fuels 128–129, 131
    France 130–131, 134–135, 149
    hydro-electric power 130, 132, 161, 178
    India 132–133, 136–137, 149
    Mohave Desert, USA 179
    nuclear power 130–131, 178, 189
    power stations 136, 144, 186, 189
    renewable 77
    solar 77, 178
    sustainable supplies 130, 134–137, 142
    thermal power 131, 134–135
    and weather 178–179
    wind power 179
environment 7–8, 23, 29, 58, 65, 72–73, 76, 98, 121, 129, 134–136, 142, 155, 165, 169–171, 178–181, 188, 190, 195–197, 198, 200, 204, 212, 214, 216, 219, 221, 240, 242
Environment Agency 11
Equator 66, 180
Eritrea, Africa (drought) 216
erosion 54–65, 69, 165, 235
    abrasion/corrasion 57
    cliff face 56–57
    cliff foot 56–57

INDEX

corrosion 57
cost of protection 64–65
hard engineering 59, 62
hydraulic pressure 56
slumping 65
soft engineering 59, 62
soil 15, 76, 205, 220
sub-aerial processes 57
weathering 57
Europe (acid rain) 187–188
   (global warming) 193
   (farming) 198
   (gm food) 212
European Union (EU)
   waste management 28
   farming 201, 215
evaporation 90, 163, 171, 181
Everglades, USA (global warming) 172–173
extinction 68, 76

F

Fair Labour Association 41
Fairstead Hall Farm, UK 200
famine 162, 216
FAO (Food and Agricultural Organization) 156
farming 14–15, 32–33, 68, 72–74, 128, 139–140, 142, 155–157, 160–164, 166–168, 170, 180, 182–183, 187, 193, 195, 198–221, 232, 236, 240–243
   alternative methods 206–209
   Aviaries Farm, UK (organic farm) 206–207
   changes to 200–201
   and the environment 204, 214
   Europe 198
   Fairstead Hall Farm, UK 200
   genetic pollution 213
   intensive 213
   land degradation 216, 220
   Lynford House Farm, UK (agro-business) 202–206
   organic 72, 206–209
   Sahel, Africa 214–221
   seed cross-breeding 218–219
   subsistence 215, 220
   sustainable development 214, 221
   UK 199, 201, 204, 208
   United States 182, 185, 211
favelas (shantytowns, São Paulo) 18–20
fertilizer 206, 215
fetch 56
flood bank/walls 100
flood plains 11, 91, 100–101, 165, 171
flooding 11, 15, 18, 90–101, 162, 174, 180, 192, 195
   Africa 98–99, 101
   Bangladesh 99, 101
   causes 90–91, 98
   control of 100–101
   Cyclone Eline 98
   effects 94–97
   human factors 91
   Mozambique 96–100
   physical factors 90–91
   response to 99–100
   River Derwent, UK 92–95, 101
   River Nile, Egypt 162
Florida, USA (tourism) 176
fold mountains 80
food production (see farming)
food surpluses 213
foot and mouth disease 198–199, 232
footloose industries 52
Ford Motor Co. 34–37
fossil fuels 178, 186, 188–189, 195
France (energy) 130–131, 134–135, 149
Frankfurt Stock Exchange (resources)128
Fulani people, Africa 214, 216

G

GA (Global Alliance for Workers and Communities) 41
gabions 58
Gambia (tourism) 176
genetically modified food 182, 210–213
   advantages 211
   disadvantages 212–213
geology 57–58, 65, 226, 228, 230
   aquifers 150, 172
   impermeable rock 57–58, 91
   mass movement 57
   permeable rocks 57–58
   sedimentary rocks 150
   strata 58, 65
Germany (population) 106, 116–118, 122
   Black Forest (acid rain) 188
Ghana (subsistence farming) 33
glaciers 230
GlaxoSmithKline (TNC) 35
global warming 129, 144, 172–174, 187, 190–195, 217
   Bangladesh 193
   Bangkok 192
   China 187, 194
   Dhaka 192
   Everglades, USA 172–173
   Europe 193
   The Hague 194–195
   India 187, 194
   Kyoto summit 194
   Netherlands 192
   predictions 192
   prevention 194
Gorbachev, Mikhail 219
governments 6, 9, 11, 31, 54, 66, 68, 119, 142, 144–145, 166, 187, 189, 198, 226, 232, 240
Grand Canyon, USA (National Park) 240
green corridor 12
Green Cross programme 219
greenfield sites 6, 8, 10, 24
   advantages 10
greenhouse effect 191
greenhouse gases 129, 178, 191, 193–195
Greenpeace 212
gross domestic product/national product (GDP/GNP) 32, 101
groynes 59, 61, 64
Gujarat State, India (earthquake) 78
Gulf Stream (global warming) 194

H

Hague, The (global warming) 194–195
hazards 15, 78–101
Heddon-on-the-Wall, UK (foot and mouth) 198
Hengistbury Head, UK (coastal erosion) 56
Himalayas 81
Hoburn Naish, UK (coastal protection) 61, 65
Hoover Dam, USA 161
'Household Growth' green paper 6
housing 6–12, 15, 18–22, 71, 142, 144, 234
   Ashford, UK 8–12
   economic impact 7–8
   environmental impact 7–8, 11
   São Paulo, Brazil 16–21, 31
   social impact 7–8
human rights 41
hunting 244–5
Hurst Castle Spit, UK (coastal protection) 58, 60, 62–63
hydrographs 93

I

Ice Ages 190, 231
Iceland (plates) 80
   (climate) 182
immigration 6, 117
India (resources) 126, 128,
   (energy) 132–133, 136–137, 149
   (monsoon) 174
   (acid rain) 187
   (global warming) 194
industrialisation 158–159, 163, 187
industry 20, 32, 42–5, 48–51, 72–74, 139–140, 144, 155, 160–161, 167–168, 186, 202
   coal mining 189
   footloose 52
   hi-tech 42–43, 52–53
   service-based 30, 42
infiltration 90
irrigation 156–157, 163, 167–168, 172, 182–183, 195, 218
   Mali, Africa 218–219
   Sinai 157
Israel (water demand) 166–167

K

Kendal, UK (tourism) 228
Kosovo (migration) 117
Kyoto summit (global warming) 194–195

L

lakes 150, 187
   Lake Chad, Africa 173
   Coniston, UK 241
   Lake District, UK 175, 224–241
   Grasmere, UK 231
   Lake Nasser, Egypt 163, 165
   Thirlmere, UK 230
   Windermere, UK 228, 238–9, landforms 226
Las Vegas, USA (water demand) 160–161
LEDCs (Less Economically Developed Countries)
   climate 184
   employment 35, 38
   energy 132, 138–139
   farming 213–214
   flooding 100
   greenhouse gases 195
   population 106, 109, 112, 118, 120
   resources 146,
   tourism 242
   urbanisation 14–17, 24
   unemployment 19–22
   water 154, 158, 165, 168
life expectancy 110–111, 122
longshore drift 60–1

Index

London, UK
    (the City) 32
    (environmental sustainability) 29
    (waste management) 28–29
Los Angeles, USA (water demand) 161
Lynford House Farm, UK (agro-business) 202–206

M

M4 corridor 52–53
MAFF (Ministry for Agriculture, Forestry and Fisheries) 62
Mahindra Ltd. 36
Malawi, Africa (population) 106, 112–115, 118–121
Mali, Africa (irrigation) 218–219
malnutrition 213
Malthus, Thomas 139–140
manufacturing 30, 32, 34–37, 42–43
Mauritania, Africa (drought) 217
mechanisation 200, 204
MEDCs (More Economically Developed Countries)
    employment 35, 38
    energy 130
    farming 213–214
    flooding 100–101
    population 106, 109, 116, 118, 120, 122–123
    resources 140
    urbanisation 14, 16, 160
    water 168
Mediterranean (rainfall) 152
    (climate) 174, 193
Melchett, Lord 212
microclimates 181, 183–185
Mid-Atlantic Ridge 80
migration 15–16, 18, 20, 114, 117, 158, 214–215
    Bosnia 117
    Kosovo 117
    emigration 6
    immigration 6, 117
    pull factors 15, 18
    push factors 15, 18
milk quotas 201
Milton-On-Sea, UK (coastal erosion) 55, 58, 62
Mohave Desert, USA (wind power) 179
Monsanto 210, 212
monsoon 174
Montserrat (volcano) 78, 80, 82–83
More, Sir Thomas 22
Mount Pelee, Martinique (volcano) 83
Mount Pinatubo, Philippines (volcano) 78, 85
Mount St Helens, USA (volcano) 80
Mozambique (floods) 96–100
Mumbai, India (global warming) 192

N

national insurance 31
National Park Authority 233, 240
National Parks
    Donana, Spain 173
    Canada 226
    Grand Canyon, USA 240
    Lake District, UK 175, 224–241
    Sahel, Africa 214–221
    Yosemite, USA 240
    Zimbabwe 242–5
National Trust 225
natural disasters (see hazards)

Netherlands (global warming) 192
New Zealand (drought) 178
NIC (newly industrializing country) 32
Niger, Africa (drought) 217
Nike 38–41
    working conditions 39–41
NIMBYism 8, 179
Nogent-sur-Seine Power Station, France 131
North Pole (climate) 182
North Sea (flooding) 92–93
North York Moors (flooding) 92–93
Norway (acid rain) 189

O

oceanic crust 80
off shore breaks 59
ozone layer 188, 190

P

Phoenix, USA (water demand) 160–161
Plate tectonics 79
plates 78–81
    collision margins 81
    conservative margins 81
    constructive margins 80
    continental 79
    destructive margins 80
    Eurasian Plate 81
    Himalayas 81
    Indian Plate 81
    Nazca Plate 80
    North American Plate 81
    oceanic 79
    Pacific Plate 81
    plate margins 80–81
    South American Plate 80
pollution 7, 11, 18, 24–25, 27, 50, 76, 135–136, 140, 142, 154, 168–169, 171–172, 178, 186–187, 189, 239
population 102–125
    ageing 6, 117–118, 122–125
    birth rate 14, 102–103, 110, 112, 116, 120, 122
    cohorts 112
    China 102, 104–105, 146–147, 158–159
    death rate 14, 102–103, 110, 116
    decreases 107, 109
    dynamics 102–125
    economically active 117
    Egypt 156–157
    Germany 106, 116–118, 122
    growth 6, 14–16, 36, 77, 102–104, 106–110, 112, 122, 139–140, 148, 156–159, 163, 170, 172, 213, 215
    growth rate 14, 103, 106–108, 110, 122
    Malawi 106, 112–116, 118–121
    models 106, 111
    natural increase 102–103
    patterns of change 106–108, 110
    prediction 107, 112
    processes 116
    pyramids 112, 116, 122
    and resources 138–149
    structure 112
Port Douglas, Australia (coastal development) 70–71, 76
poverty 119, 220, 243
precipitation 150, 178
Projet Agro-Forestier 220

R

rainfall 68, 98–99, 151–154, 163, 165, 173–174, 178, 180, 183, 185, 202, 206, 215–216, 218–220
Rainforest Co-operative Research Centre 76
rainforests 83, 128, 190
    Daintree, Australia 66, 68–69, 71–74, 76–77
Reading (employment) 42–51
    Football Club 46–47
    out of town shopping 48–51
recreation 222–245
recycling 143, 145
refugees 114–115
reservoirs 100, 150–151, 153, 241
resettlement 165
resource stock 134
resources 76, 126–149, 221
    change in demand 139
    non-renewable 126–129, 135, 142
    predicting demand 139, 149
    renewable 76, 126–129, 135, 142
    substitution 129
    sustainable 130, 134–137, 142, 146, 168, 173
    water 150, 173
    weather 174, 196
revetments 58, 64
Richter Scale 78, 86
rivers 100, 150–151, 168, 172, 187
    River Colorado 160–161, 172
    River Daintree 75–76
    River Danube 168–169
    River Derwent, Yorkshire 92–95
    Huang He River 172
    Huangpo River 159
    River Mississippi 172
    River Nile 156–157, 162–165
    River Tisza 168
    Thames 190
    Yangtze 172
    Zambezi 242
runoff 90
rural areas 222–245
    development of 7, 10, 15, 242
    impact of development 7

S

Sahara Desert 216
Sahel, Africa (farming) 214–221
salinisation 165, 171
salt marshes 60, 62
San Andreas Fault, USA 81, 89
São Paulo, Brazil (settlement) 16–21, 31
sea levels 192
seismographs 88
seismometers 84
Sellafield 189
services
    education 119
    electricity 18–21, 75
    health 111, 119, 243
    medical 14–15, 18–20
    sanitation 14–15, 18–21, 75, 111, 159
    schools 8, 10, 15, 18–20, 243
    service-providers 30
    water 75, 243
settlement 6–29
    Ashford, UK 8–12
    São Paulo, Brazil 16–21, 31

INDEX

sustainability 6–9
Shanghai, China (water demand) 159
Siguin Vousse, Africa (tourism) 220–221
Sinai (irrigation) 157
snow 177–178
South Africa (flood) 98–99
    (population) 111
South East Asia (employment) 35, 37–41
South Korea (employment) 32, 38
Soviet Union (water management) 170
spits 60, 62–63
Spitzbergen Syndrome 194
SSSI (Site of Special Scientific Interest) 62
Sudan (water management) 163
Sunungukai Camp 243
supermarkets 49, 201, 208
Suzhou, China (water demand) 159

T

tax 31, 54, 69, 123, 195
TCPA (Town and Country Planning Association) 6
technology 183, 195
telemetry stations 101
throughflow 90
TNCs (transnational corporations) 30, 35, 42, 190
tourism 65–76, 83, 161, 175–177, 222–245
    advantages 232–233, 242–243
    Australia 74, 76–77, 226
    Cornwall, UK 224–225
    Dubai 176
    disadvantages 234–235
    ecotourism 242–245
    Gambia 176
    impact 232, 234–237, 240, 243
    Kendal, UK 228
    Lake District 224–241
    Siguin Vousse, Africa 220–221
    sustainable development 236–245
    UK 222–223
    Victoria Falls, Africa 242
    Zimbabwe 242–245
town planning 6, 7, 12, 23, 25
transport 8, 10, 12, 15, 19, 20, 24–27, 43, 49, 67, 74–75, 142, 144, 195, 225, 234–235, 240
    Birmingham, UK 24–25
    Boston, USA 26–27
    roads 10, 18–19, 26–27, 46–49, 52–53
    sustainability 25–27
transport corridors 6
Tucson, USA (water demand) 160–1

U

unemployment 8, 18–19, 42, 23–25
United Kingdom
    (acid rain) 189, 193–194
    (climate) 174–175
    (drought) 153
    (employment) 30–31
    (farming) 199, 201, 204, 208
    (flooding) 90
    (foot and mouth disease) 198
    (gm food) 212–213
    (housing) 6–7, 12
    (tourism) 222–223
    (sustainable development) 22
    (water) 150–151
United Nations 114, 120, 154, 192

United States of America
    (farming) 182, 185, 211
    (greenhouse gases) 195
    (National Parks) 226, 240
    (urbanisation) 160
urban areas 155, 158
    climates 184–185
    water 155, 158
urban growth/urbanisation 6–7, 11, 13–29, 43, 49, 101, 158, 160, 163, 184
    effects 18
    sustainability 8, 22–29
urban heat islands 184–185
urban regeneration 6, 20–21

V

Vesuvius, Italy (volcano) 85
Victoria Falls, Africa (tourism) 242
volcanoes 78–85, 88, 191
    ash 82
    Chances Peak, Montserrat 82
    effects 82–84
    fissure eruptions 80
    lava 80, 82
    magma/molten rock 80, 84, 88
    Montserrat 78, 80, 82–83
    Mount Pinatubo, Philippines 78, 85
    Mount St Helens, USA 80
    Mount Pelee, Martinique 83
    patterns 78–79
    planning 85
    predicting 84–85
    pyroclastic flow 82–83, 85
    subduction zone 80
    Vesuvius, Italy 85

W

wars 114
Washington State, USA (earthquake) 78
waste management 7, 20, 28–29, 46, 142–143
    environmental impact 28
    sustainability 28
water 14, 18–21, 150–173
    Aral Sea 170–171
    Aswan Dams 162–165
    Beijing, China 172
    Belgrade, Serbia 168
    Cherrapunji, India 152
    consumption 154
    Donana, Spain 173
    Egypt 162–165
    management 169, 173
    demand 156, 159, 172–173
    demand management 166, 173
    groundwater supplies 151, 153, 187
    Israel 166–167
    quality 165, 169
    as resource 150, 173
    salinisation 165, 171
    sources 150–151
    Shanghai, China 159
    Soviet Union 170
    stakeholders 167
    Sudan 163
    sustainable use 168–169
    Suzhou, China 159
    USA 160–161
    varying supplies 150–154
    virtual water 166

    waste 168
    Yorkshire, UK 150–151, 153
water cycle 150
water stress 154
water table 153
waves 56–58, 60–61, 65
weather 174–197, 216
    control of 174
    and energy 178
    modification 182
    as resource 174, 196
wetlands 172–173
World Bank 120, 136
World Heritage Sites
    Daintree, Australia 68, 76
    Danube 169

Y

Yorkshire Wolds, UK (water supply) 150–151, 153
Yosemite, USA (National Park) 240

Z

Zimbabwe (ecotourism) 242–245